Björn Dollmann

Dynamic Nuclear Polarization

Björn Dollmann

Dynamic Nuclear Polarization

Advances in Hardware and Molecular
Design of Polarizing Agents

Südwestdeutscher Verlag für Hochschulschriften

Impressum/Imprint (nur für Deutschland/ only for Germany)
Bibliografische Information der Deutschen Nationalbibliothek: Die Deutsche Nationalbibliothek verzeichnet diese Publikation in der Deutschen Nationalbibliografie; detaillierte bibliografische Daten sind im Internet über http://dnb.d-nb.de abrufbar.

Alle in diesem Buch genannten Marken und Produktnamen unterliegen warenzeichen-, marken- oder patentrechtlichem Schutz bzw. sind Warenzeichen oder eingetragene Warenzeichen der jeweiligen Inhaber. Die Wiedergabe von Marken, Produktnamen, Gebrauchsnamen, Handelsnamen, Warenbezeichnungen u.s.w. in diesem Werk berechtigt auch ohne besondere Kennzeichnung nicht zu der Annahme, dass solche Namen im Sinne der Warenzeichen- und Markenschutzgesetzgebung als frei zu betrachten wären und daher von jedermann benutzt werden dürften.

Verlag: Südwestdeutscher Verlag für Hochschulschriften GmbH & Co. KG
Dudweiler Landstr. 99, 66123 Saarbrücken, Deutschland
Telefon +49 681 37 20 271-1, Telefax +49 681 37 20 271-0
Email: info@svh-verlag.de
Zugl.: Mainz, Johannes-Gutenberg Universität, Dissertation, 2011

Herstellung in Deutschland:
Schaltungsdienst Lange o.H.G., Berlin
Books on Demand GmbH, Norderstedt
Reha GmbH, Saarbrücken
Amazon Distribution GmbH, Leipzig
ISBN: 978-3-8381-2560-2

Imprint (only for USA, GB)
Bibliographic information published by the Deutsche Nationalbibliothek: The Deutsche Nationalbibliothek lists this publication in the Deutsche Nationalbibliografie; detailed bibliographic data are available in the Internet at http://dnb.d-nb.de.

Any brand names and product names mentioned in this book are subject to trademark, brand or patent protection and are trademarks or registered trademarks of their respective holders. The use of brand names, product names, common names, trade names, product descriptions etc. even without a particular marking in this works is in no way to be construed to mean that such names may be regarded as unrestricted in respect of trademark and brand protection legislation and could thus be used by anyone.

Publisher: Südwestdeutscher Verlag für Hochschulschriften GmbH & Co. KG
Dudweiler Landstr. 99, 66123 Saarbrücken, Germany
Phone +49 681 37 20 271-1, Fax +49 681 37 20 271-0
Email: info@svh-verlag.de

Printed in the U.S.A.
Printed in the U.K. by (see last page)
ISBN: 978-3-8381-2560-2

Copyright © 2011 by the author and Südwestdeutscher Verlag für Hochschulschriften GmbH & Co. KG and licensors
All rights reserved. Saarbrücken 2011

To Nina

Contents

1. **Introduction**   1

2. **Theoretical Background**   3
   - 2.1. NMR Fundamentals 3
   - 2.2. EPR Fundamentals 13
   - 2.3. Dynamic Nuclear Polarization 21

3. **Technical Aspects of the Mobile Set-up**   41
   - 3.1. Experimental Set-up 41
   - 3.2. Halbach Magnet versus Electromagnet 42
   - 3.3. Shimming of a Halbach Magnet 47
   - 3.4. DNP Probeheads 51
   - 3.5. Implementation of LabVIEW 56

4. **DNP Performance of the Probeheads**   59
   - 4.1. CUBOID Probehead 59
   - 4.2. PH1004 Probehead 62
   - 4.3. Summary 63

5. **Overhauser-type DNP Performance of Polarizing Agents**   65
   - 5.1. TEMPOL 65
   - 5.2. Spin-Labeled Heparin 73
   - 5.3. Thermoresponsive Spin-Labeled Hydrogel 87
   - 5.4. Summary - Polarizing Agents 95

6. **Solid-State DNP Performance of Polarizing Agents**   97
   - 6.1. TEMPOL 99
   - 6.2. Spin-Labeled Heparin 106
   - 6.3. Thermoresponsive Spin-Labeled Hydrogel 118
   - 6.4. Summary - Solid-State DNP 120

7. Hyperpolarization of Hetero Nuclei — 123
  7.1. Hexafluorobenzene 125
  7.2. DNP of a Dissolved ^{19}F Containing Molecule 131
  7.3. ^{13}C-Enriched Urea 133
  7.4. Summary - Hyperpolarization of Hetero Nuclei 139

8. Conclusion — 141

Appendix — 147

A. Appendix - Methods — 147
  A.1. Determination of Unknown Radical Concentrations 147
  A.2. CW EPR Line Shape Analysis 147
  A.3. ESE-Detected Line Shape Analysis 148
  A.4. DEER Analysis 149
  A.5. Electron spin-lattice determination at room temperature 149
  A.6. DNP Analysis 150
  A.7. Determination of the Quality Factor of the EPR Probeheads 151

B. Materials - Polarizing Agents, Solvents and Used Molecules — 154
  B.1. TEMPO Derivatives 154
  B.2. Spin-Labeled Heparin 155
  B.3. SL-Hydrogel 159
  B.4. Solvents 161
  B.5. Solutes 162

C. Matlab Scripts — 163
  C.1. Enhancement and Power Dependence 163
  C.2. Magnetic Field Sweep 169

Bibliography — 174

1. Introduction

Nuclear magnetic resonance (NMR) is a versatile technique relying on spin-bearing nuclei. Since its discovery more than 60 years ago,[1,2] NMR and related techniques have become indispensable tools with innumerable applications in physics, chemistry, biology and medicine. One of the main obstacles in NMR is its notorious lack of sensitivity, which is due to the minuscule energy splitting caused by the nuclear spins at room temperature. Even for proton spins, which possess the largest magnetogyric ratio, the degree of polarization in the highest available magnetic fields (24 T) is only $\sim 7 \cdot 10^{-5}$. Appropriately, the inherent low polarization allows for a theoretical sensitivity enhancement of more than 10^4. In the field of magnetic resonance imaging (MRI) this issue becomes even more severe as the magnetic fields of whole-body tomographs do not reach the field strengths of magnets for NMR spectroscopy. Accordingly, MRI is mainly restricted to the imaging of water protons and the application of ^{13}C (or other low γ nuclei) NMR spectroscopy and imaging for clinical diagnosis has been constrained by the extremely long imaging and spectroscopy acquisition times that are required to obtain high signal-to-noise ratios under physiological conditions (low natural abundance of ^{13}C and low concentration of ^{13}C-compounds).

Due to this potential sensitivity increase which can open up completely new research fields in NMR spectroscopy and imaging, several hyperpolarization techniques have been developed to overcome this drawback of NMR. The hyperpolarization techniques can be divided into two sub-groups: (i) Chemistry-based polarization methods like *e.g.* para-hydrogen induced polarization[3–5] (PHIP) and photochemically induced dynamic nuclear polarization[6–8] (Photo-CIDNP). (ii) Physics-oriented polarization methods like *e.g.* optical pumping of noble gases[9,10] and dynamic nuclear polarization[11–15] (DNP) which was used in this thesis.

DNP is a polarization technique which transfers the polarization of unpaired electron spins to the surrounding nuclei by microwave irradiation. The theoretical enhancement is given by the ratio of the magnetogyric ratios of the electron spin and a nuclear spin. This ratio for proton (^1H), fluorine (^{19}F) and carbon (^{13}C) nuclei is $E \approx 660$, $E \approx 700$ and $E \approx 2600$, respectively. The most characteristic feature and significant drawback are the presence of unpaired electrons in the sample which have to be added if they are not inherent to the sample. On the other hand, DNP is not limited to a special molecule or nucleus which makes it remarkably versatile.

With respect to the possible NMR signal enhancements, especially for low γ nuclei, many new experiments in physics, chemistry, biology and medicine emerge. Particularly in medicine new diagnostic pathways can be taken by utilizing hyperpolarized substances as active contrast agent.[13,16] For example, the metabolism of the physiological relevant substance pyruvate could be examined via MRI.[16,17] Thus, further development of technical components and suitable polarizing agents for Overhauser-type and solid-state DNP are of high importance.

So far, only few medical applications are based on the DNP technique. Nevertheless, this thesis deals exclusively with the technical development of a mobile DNP polarizer and the design of suitable polarizing agents for DNP and prospective medical applications. After this introductory Chapter, the basic theoretical background is given to comprehend the experimental results of the following parts. The first part will discuss the technical improvements which could be achieved regarding the mobile DNP polarizer. These improvements comprise the realization of a homogeneous Halbach magnet, the implementation of an automated experiment control and the construction of new probeheads. In the Chapters 3 and 4 the importance of these technical improvements will be demonstrated.

In the main part of this thesis the DNP performance of new polarizing agents is presented and compared to a commonly used polarizing agent. The DNP performance was tested at physiological and cryogenic temperatures. Appropriately, Chapter 5 deals with the Overhauser-type DNP and Chapter 6 with the solid-state DNP. These two Chapters particularly focus on the biocompatibility and removal of polarizing agents which is an important issue concerning the medical application of hyperpolarized substances via DNP.

In Chapter 7 the feasibility of DNP experiments on hetero nuclei in the mobile set-up, working at 0.35 T, is demonstrated. Along with the achieved NMR signal enhancements by DNP, the measurements prove the possibility of a fast and reliable NMR detection and nuclear spin-lattice relaxation time determination of biomolecules by using DNP. This experiment shows the potential of DNP, especially at X-band, when it comes to the polarization of nuclei with a very low magnetogyric ratio.

2. Theoretical Background

This Chapter introduces the basic knowledge to understand and interpret the data and results summarized in the thesis. The presented theoretical fundamentals comprise the topics introduced and explained in many publications, books and reviews.[11,12,18] The Chapter can be divided into four parts. First, the basic principles of nuclear magnetic resonance (NMR) are explained and its limitations are discussed. Next, the most important properties of the similar method electron paramagnetic resonance (EPR) and its advantages and disadvantages as compared to NMR are presented. In Section 2.3 the basics of dynamic nuclear polarization (DNP) are described in the context of the benefits of combining NMR and EPR and its use for new and powerful applications, especially in the wide field of medicine. The last part concludes with the used theoretical tools for the shimming of a Halbach magnet by utilizing permanent magnets.

2.1. NMR Fundamentals

NMR and magnetic resonance in general rely on the inherent quantum mechanical property of elementary particles: the *spin*. Spin-bearing particles possess a non-zero spin angular momentum which is an intrinsic property, like the mass, of the particle. Neutrons and protons which are the constituents of nuclei carry the spin-quantum number 1/2. Therefore, the resulting spin-quantum number of a nucleus can be either integer or half-integer. Its characteristic magnetic moment μ_I for a non-zero spin-quantum number is collinear to its momentum and defined by

$$\boldsymbol{\mu_I} = \gamma_I \hbar \boldsymbol{I} \; , \tag{2.1}$$

where γ_I denotes the magnetogyric ratio of the nucleus and $\hbar = h/2\pi$ is Planck's constant divided by 2π. The time evolution of observables in quantum mechanics (QM) obeys the time-dependent Schrödinger Equation 2.2:

$$i\hbar \frac{\partial}{\partial t} |\Psi(t)\rangle = \hat{\mathcal{H}} |\Psi(t)\rangle \quad , \tag{2.2}$$

where $|\Psi(t)\rangle$ is the state of the spin system at time t and $\hat{\mathcal{H}}$ is the Hamilton-operator (Hamiltonian). The spin Hamiltonian of a nuclear system at a magnetic field reads as

$$\hat{\mathcal{H}} = \hat{\mathcal{Z}}_I + \hat{\mathcal{H}}_{rf} + \hat{\mathcal{H}}_{cs} + \hat{\mathcal{H}}_{II} + \hat{\mathcal{H}}_Q \quad . \tag{2.3}$$

In 2.3 $\hat{\mathcal{Z}}_I$ is the nuclear Zeeman interaction with the externally applied magnetic field and $\hat{\mathcal{H}}_{rf}$ represents the interaction of the nuclear spins with the radio-frequency (rf) field. $\hat{\mathcal{H}}_{cs}$ denotes the interaction due to chemical shift. This term is omitted as the spectral resolution and its analysis are of minor interest for the purpose of the thesis. The last two terms represent the mutual interactions due to dipolar and quadrupolar interactions, respectively. Quadrupolar interactions emerge only for nuclear spins with $I > 1/2$. As this work exclusively deals with $I = 1/2$ nuclei (^{1}H and ^{13}C), it is not further considered here. Furthermore, as for DNP all samples are doped with radicals (unpaired electrons), the nuclear spin-nuclear spin interactions are at least one order of magnitude smaller than electron spin-nuclear spin interactions.[18,19] Therefore, the only remaining interactions are of external kind and will be discussed in the following Subsections.

External Nuclear Spin Interactions

$\hat{\mathcal{Z}}_I$ in Equation 2.3 is the interaction energy of the nuclear spin momentum with the applied external magnetic field $\boldsymbol{B}_0$:

$$E = \hat{\mathcal{Z}}_I = -\boldsymbol{\mu}_I \boldsymbol{B}_0 = -\gamma_I \hbar \boldsymbol{I} \cdot \boldsymbol{B}_0 \quad . \tag{2.4}$$

If the external magnetic field $\boldsymbol{B}_0$ is aligned along the z-axis in the laboratory frame, $\boldsymbol{B}_0 = (0, 0, B_0)^T$, the energy levels depend only on the magnetic spin-quantum number

2. Theoretical Background

m:

$$E_m = -\gamma_I m \hbar B_0 \ . \tag{2.5}$$

The magnetic spin-quantum number m can only acquire the values $|m| \leq I$. Every state with the spin-quantum number m can occupy $2m+1$ sub-levels which are degenerate in the absence of an external magnetic field. The energy difference between these sub-levels scales linearly with the applied magnetic field:

$$\Delta E = E_{m+1} - E_m = \gamma_I \hbar B_0 = \hbar \omega_0 \ , \tag{2.6}$$

where the Larmor frequency $\omega_0 = -\gamma B_0$ is introduced. The Larmor frequency determines the resonance frequency at which the spins precess around the vector of the B_0 field. For example, the Larmor frequency of proton spins with $I = 1/2$ ($\gamma_I = 42.58$ MHz/T) at a magnetic field of $B_0 = 0.35$ T leads to a precession frequency of $f_0 = \omega_0/2\pi \approx 14.9$ MHz corresponding to an energy difference $\Delta E = \hbar \omega_0 \approx 3 \cdot 10^{-7}$ eV. This value of the cyclic frequency corresponds to the radio-frequency (rf) spectrum. From a classical point of view, this phenomenon corresponds to the precession of a nuclear angular moment around the static field.

In NMR the observed experimental system consists of an ensemble of atoms ($\sim 10^{23}$ spins), which cannot be described with a quantum-mechanical wave-function. As NMR experiments are performed at ensembles with almost identical properties/systems, a macroscopic magnetization $\boldsymbol{M}$ can be introduced which represents the sum of all microscopic spin orientations. This simplification leads to a more intuitive interpretation of NMR experiments. The energy splitting calculated in Equation 2.6 implies a difference in the population of the states N_+ and N_- which obeys the Boltzmann distribution (Figure 2.1)

$$\frac{N_+}{N_-} = \exp\left(-\frac{\Delta E}{k_B T}\right) = \exp\left(-\frac{\gamma_I \hbar B_0}{k_B T}\right) \ , \tag{2.7}$$

where T denotes the absolute temperature and k_B the Boltzmann constant.
The excess population which is responsible for the resonance phenomenon is defined by

2. Theoretical Background

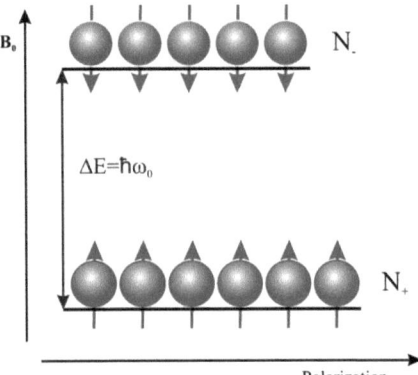

Figure 2.1.: Population of a spin-$\frac{1}{2}$ system with a positive magnetogyric ratio as it is the case for hydrogen and carbon nuclei. The population of the energy levels obeys the Boltzmann distribution.

the *population P*

$$P = \frac{N_+ - N_-}{N_+ + N_-} = \frac{1 - \exp\left(-\frac{\gamma_I \hbar B_0}{k_B T}\right)}{1 + \exp\left(-\frac{\gamma_I \hbar B_0}{k_B T}\right)} \quad . \tag{2.8}$$

In the "high-temperature approximation" which is valid for $T > 2$ K Equation 2.7 can be expanded

$$P = \tanh\left(\frac{\hbar \gamma_I B_0}{2 k_B T}\right) \approx \frac{\hbar \gamma_I B_0}{2 k_B T} \quad . \tag{2.9}$$

For a typical water sample at room temperature and a static field of $B_0 = 0.35$ T, the polarization of ^{1}H is only $P = 1.2 \cdot 10^{-6}$, which justifies the validity of the approximation. The total magnetization $\boldsymbol{M}_0 = (0, 0, M_0)^T$ for $\boldsymbol{B}_0 = (0, 0, B_0)^T$ at thermal equilibrium is then given by the sum of all nuclear magnetic moments

$$\boldsymbol{M}_0 = \sum_i^N \boldsymbol{\mu}_i = \frac{1}{2} N \hbar \gamma_I P \hat{e}_z \quad , \tag{2.10}$$

where $N = N_+ + N_-$ is the total number of nuclear spins.

The characteristic energy splitting of an atomic nuclear spin leads to a specific resonant absorption and dispersion line, if an oscillating magnetic field of amplitude B_1 is applied transverse to a set of equivalent spins in thermal equilibrium in a longitudinal field $B_0\,\hat{e}_z$. The Hamiltonian in the laboratory frame can then be written as the sum of a static and an oscillatory part

$$\hat{\mathcal{H}}_{lab} = \gamma_I B_0 \hat{I}_z - 2\gamma_I B_1 \cos(\omega t)\hat{I}_x = \omega_0 \hat{I}_z - 2\omega_1 \cos(\omega t)\hat{I}_x \ , \qquad (2.11)$$

where the frequency of the transverse magnetic field $\omega_1 = -\gamma_I B_1$ was introduced. In a simple vector model the linearly polarized oscillatory field can be represented as a sum of two counter-rotating circularly polarized components with the same amplitude B_1 as shown in Figure 2.2.

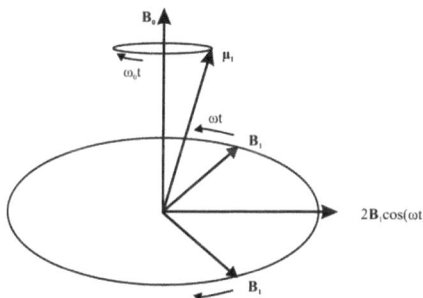

Figure 2.2.: Representation of linearly polarized, oscillatory transverse field $2B_1 \cos(\omega t)$ as two counter-rotating circularly polarized fields. The precession of the magnetic moment around the B_0 field is shown for the case of vanishing transverse magnetization.

This picture visualizes that one component rotates in the same sense as the nuclear spin precession which causes the resonance phenomenon at the Larmor frequency $\omega = \omega_0$. The other component does not interact with the nuclear spin and can be omitted for $B_1 \gg B_0$. It is profitable to transform the spin operators in the Hamiltonian (Equation 2.11) to a frame of reference which rotates with the radio-frequency (rf) ω_{1e}. In the rotating frame the Hamiltonian of Equation 2.11 is transformed to

$$\hat{\mathcal{H}}_{rot} = (\omega_0 - \omega)\hat{I}_z - \omega_{1e}\hat{I}_x \ . \qquad (2.12)$$

Thereby, the apparent precession frequency of the first term is reduced by the radio-frequency. Obviously, at the resonance condition $\omega = \omega_0$, the longitudinal field vanishes and the total magnetic moment starts rotating around the remaining transverse perturbation field. This phenomenon allows for a manipulation of the magnetization orientation. It is expressed by

$$\hat{\mathcal{H}}_{rot} \approx \hat{\mathcal{H}}_{rf} = \omega_1 \hat{I}_x = \frac{1}{2}\omega_1(\hat{I}_+ + \hat{I}_-) \ , \qquad (2.13)$$

where only the perturbation remains and the step operators $\hat{I}_\pm = \hat{I}_x \pm i\hat{I}_y$ were introduced.

The Semi-Classical Description and Relaxation Processes

Starting from the Bloch Equations which treat the motion of the total magnetization vector in a semi-classical way, one can visualize the impact of a radio-frequency perturbation on the orientation of the magnetization for a spin-$\frac{1}{2}$ system. A derivation of the Bloch Equations starts from the dynamics of the total magnetization

$$\frac{d\boldsymbol{M}}{dt} = \gamma_I \boldsymbol{M} \times \boldsymbol{B} \ , \qquad (2.14)$$

where $\boldsymbol{B} = (B_1\cos(\omega_0 t), B_1\sin(\omega_0 t), B_0)^T$ denotes the superposition of the static and the oscillating magnetic field yields

$$\begin{aligned}
\frac{dM_x}{dt} &= \gamma_I \left[M_y B_0 + M_z B_1 \sin(\omega_0 t) \right] \\
\frac{dM_y}{dt} &= \gamma_I \left[M_z B_1 \cos(\omega_0 t) - M_x B_0 \right] \\
\frac{dM_z}{dt} &= \gamma_I \left[-M_x B_1 \sin(\omega_0 t) - M_y B_1 \cos(\omega_0 t) \right]
\end{aligned} \qquad (2.15)$$

and their solutions

$$\begin{aligned}M_x &= M_0 \sin(\omega_1 t) \sin(\omega_0 t) \\ M_y &= M_0 \sin(\omega_1 t) \cos(\omega_0 t) \\ M_z &= M_0 \cos(\omega_1 t) \ .\end{aligned} \qquad (2.16)$$

In the rotating frame the motion corresponds to a precession of the magnetization about the $\boldsymbol{B}_{1e}$ field when $\omega = \omega_0$. For a duration time t_p of the rf pulse, the manipulation of the magnetization is given by the *flip angle* $\alpha = \omega \cdot t_p$, where α is the resulting angle between $\boldsymbol{M}_0$ and $\boldsymbol{B}_0$. In the case of off-resonant irradiation, the magnetization vector starts rotating around an effective magnetic field $\boldsymbol{B}_\text{eff} = \boldsymbol{B}_{1e} + (B_0 - \omega_0/\gamma_I)\,\hat{e}_z$ which is a superposition of the longitudinal and transverse field in the rotating system.
In NMR, relaxation processes are added to Equation 2.15 phenomenologically:

$$\begin{aligned}\frac{dM_x}{dt} &= \gamma_I \left[M_y B_0 + M_z B_1 \sin(\omega_0 t)\right] - \frac{M_x}{T_{2n}} \\ \frac{dM_y}{dt} &= \gamma_I \left[M_z B_1 \cos(\omega_0 t) - M_x B_0\right] - \frac{M_y}{T_{2n}} \\ \frac{dM_z}{dt} &= \gamma_I \left[-M_x B_1 \sin(\omega_0 t) - M_y B_1 \cos(\omega_0 t)\right] - \frac{M_z - M_0}{T_{1n}} \ ,\end{aligned} \qquad (2.17)$$

where M_0 denotes the magnetization at thermal equilibrium. Further, the nuclear spin-lattice relaxation time T_{1n} and the nuclear spin-nuclear spin relaxation time T_{2n} were introduced which represent the longitudinal and transverse relaxation, respectively. T_{1n} is the time the nuclear spins take to align along the longitudinal magnetic field when the system is in equilibrium. The spin-lattice relaxation is caused by the interaction of vibrating and rotating nuclei with the produced magnetic field of the lattice. This interaction leads to a dissipation of the energy transfered via an rf pulse with T_{1n} which limits the repetition time to approximately five times T_{1n}. T_{2n} describes the loss of coherence of the ensemble of the nuclear spins by nuclear spin-nuclear spin interactions in the transverse plane. This kind of interaction is a pure loss of coherence in which the energy is conserved.

Free Induction Decay and NMR Spectrum

The simplest NMR experiment is the application of a $\pi/2$ rf pulse which rotates the magnetization vector into the transverse plane followed by the acquisition of the signal which decays in a characteristic exponential manner. This decaying signal is called the *free induction decay* (FID). Such an experiment is depicted in Figure 2.3.

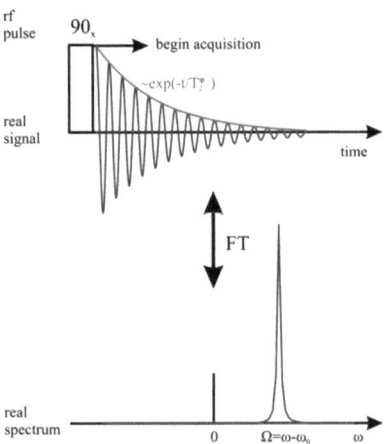

Figure 2.3.: *Free induction decay* (FID) following a 90 degree ($\pi/2$) rf pulse. Only the real part of the signal is shown which decays exponentially with the effective relaxation time T_2^*. On Fourier transformation the real signal gives an absorption spectrum at the offset frequency $\Omega = \omega - \omega_0$ with Lorentzian line shape.

The FID oscillates with the offset frequency $\Omega = \omega - \omega_0$ and decays with a relaxation time T_2^*. The relaxation time T_2^* determines the Full-Width-Half-Maximum (FWHM) of the Lorentzian NMR line which is $1/\pi T_2^*$. The effective relaxation time T_2^* consists of an irreversible part, the introduced T_{2n}, and a reversible part due to inhomogeneities of the external field and nuclear spin-nuclear spin interactions. It is approximated by

$$\frac{1}{T_2^*} = \frac{1}{T_{2n}} + \frac{1}{T'_{2n}} + \gamma_I \Delta B_0 \, , \tag{2.18}$$

where T_{2n} is referred to as homogeneous NMR line broadening and T'_{2n} and $\gamma_I \Delta B_0$ are referred to as inhomogeneous NMR line broadening. NMR experiments inherently lack sensitivity due to the low magnetogyric ratio of nuclei. Therefore, NMR experiments are repeated and accumulated for the sake of signal-to-noise (SNR) enhancement. The NMR experiment can be repeated (almost) without loss of signal after roughly five times T_{1n} as the magnetization has relaxed into its thermal equilibrium again. The nuclear spin-lattice relaxation time can be measured by an inversion recovery experiment as described elsewhere.[20,21]

Spin Echo

The *spin echo* is the re-emerging NMR signal when the reversible part of the signal decay is recovered by spin manipulations. In physics, this phenomenon can be understood as an inversion of the time evolution of the ensemble of nuclear spins meaning that the "time runs backwards". The *spin echo* was first discovered by E. Hahn in 1950[22] which lead to many new echo sequences in magnetic resonance.

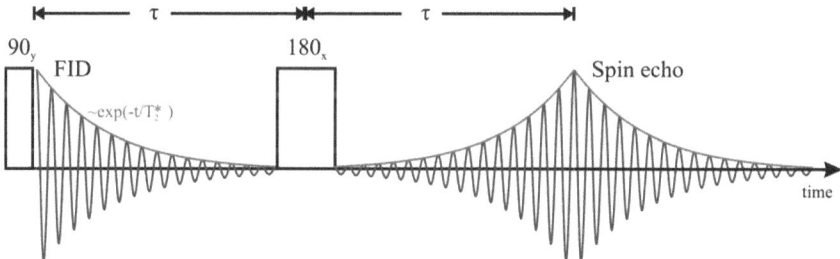

Figure 2.4.: Two pulse *spin echo* sequence, which refocuses the NMR signal at time 2τ after the 90_y rf pulse.

The spin echo sequence consists of two pulses as shown in Figure 2.4. The first pulse is a 90_y degree pulse, generating an FID signal which decays with T_2^*. After the pulse, the spin system evolves for a time τ. Within τ the nuclear spin bundles de-phase as they exhibit different Larmor frequencies, *e.g.* spin bundle A with ω_A and spin bundle B with $\omega_B = \omega_A + \Delta\omega$, resulting in a de-phasing of the spin packages at time τ: $\Delta\phi_1 = (\omega_A - \omega_B) \cdot \tau = \Delta\omega \cdot \tau$. After the time τ a second pulse (180_x) is applied which reverts the precession direction of the spin bundles. After the second pulse, spin bundle A evolves with "$-\omega_A$" and B with "$-\omega_B = -(\omega_a + \Delta\omega)$" (referred to as "running back in time"), causing a de-phasing of opposite signs: $\Delta\phi_2 = (-\omega_A - (-\omega_B)) \cdot t = -\Delta\omega \cdot t$. As $-\Delta\phi_2$ equals $\Delta\phi_1$ for $t = \tau$, the NMR signal reappears as a *spin echo* at time 2τ as it is shown in Figure 2.4. As a *spin echo* sequence refocuses only the reversible magnetization loss, the amplitude of the *spin echo* is reduced by the factor $\exp(-2\tau/T_{2n})$. Thereby, a repeated iteration of the second pulse can be used to measure the nuclear spin-nuclear spin relaxation time T_{2n}. This so-called *CPMG* pulse sequence is named after its inventors Carr, Purcell, Meiboom and Gill.[23] Supplementary, the *CPMG* pulse sequence can significantly increase the signal-to-noise ratio by accumulating in single shot experiments as many echoes as feasible.

2.2. EPR Fundamentals

EPR spectroscopy is the pendant to NMR for unpaired electron spins. Electrons are elementary particles in the atomic shell with an intrinsic spin $S = \frac{1}{2}$. They are lighter than protons by three orders of magnitude and consequently possess a magnetic moment μ_S which is by approximately three orders of magnitude higher as compared to protons:

$$\boldsymbol{\mu}_S = -g_e \mu_B \boldsymbol{S} \; , \tag{2.19}$$

where $g_e \approx 2.0023$ is the free electron *g-factor* and where the *Bohr magneton* $\mu_B = \frac{eh}{4\pi m_e}$ was introduced. The energy of a magnetic moment $\boldsymbol{\mu}_S$ is introduced analogous to Equation 2.5 in the previous Section:

$$E_S = -\boldsymbol{\mu}_S \cdot \boldsymbol{B}_0 \; . \tag{2.20}$$

For $\boldsymbol{B}_0 = B_0 \hat{e}_z$ it follows

$$E_S = g_e \mu_B m_S B_0 = m_S \hbar \omega_S \tag{2.21}$$

with the electron spin-quantum number $m_s = \pm \frac{1}{2}$ and the electron Larmor frequency $\omega_S = g_e \mu_B B_0 / \hbar$. Analogous to nuclear spins the electron spin ensemble precesses around the applied magnetic field $\boldsymbol{B}_0$. Electron spins can only align parallel ($m_S = -1/2$) or anti-parallel ($m_S = +1/2$) to the static field. Therefore, only two Zeeman levels emerge for $B_0 > 0$ as is displayed in Figure 2.5.

The population distribution obeys the Boltzmann distribution as already explained in Section 2.1. The energy splitting increases linearly with the magnetic field and can be saturated by applying an electromagnetic field of appropriate frequency ω_S. The most common magnetic field in EPR is $B_0 \approx 0.35$ T which is called X-band for historic reasons. All EPR experiments in this work were performed at X-band frequencies and fields.

2. Theoretical Background

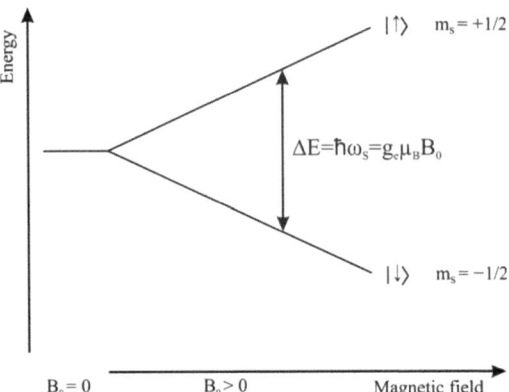

Figure 2.5.: Zeeman splitting in dependence of the applied magnetic field strength.

Interactions in EPR

The interactions of the electron spin S with an external field and with other spins are described by the following *Hamiltonian*[i]:

$$\hat{\mathcal{H}}_S = \hat{\mathcal{Z}}_S + \hat{\mathcal{H}}_{SS} + \hat{\mathcal{H}}_{SI} + \hat{\mathcal{H}}_{mw} \;. \tag{2.22}$$

The electron Zeeman interaction $\hat{\mathcal{Z}}_S$ is the dominant effect and describes the interaction of the electron spins with the external magnetic field similar to nuclear spins:

$$\hat{\mathcal{Z}}_S = \mu_B \boldsymbol{S} \cdot \underline{\boldsymbol{g}} \cdot \boldsymbol{B}_0 \;. \tag{2.23}$$

This interaction depends strongly on the symmetry of the binding orbitals or ligands. Therefore, the *free electron g-factor* from Equation 2.21 will be replaced by its anisotropic tensor $\underline{\boldsymbol{g}}$, which represents the interaction of the electron spins with the surrounding.

The electron spin-spin interaction $\hat{\mathcal{H}}_{SS}$ designates the interaction between electron spins which can be subdivided into a dipolar ($\hat{\mathcal{H}}_{dd}$) and a scalar part ($\hat{\mathcal{H}}_J$). These

---

[i]The quadrupolar interaction $\hat{\mathcal{H}}_Q$ occurs for nuclei with $|I| > 1/2$, but is of no further interest for the interpretation of the results of the thesis and is hence neglected.

2. Theoretical Background

interactions are independent of the applied magnetic field. The electron dipole-dipole coupling $\hat{\mathcal{H}}_{dd}$ of two like spins depends on the angle of the connection vector θ to the static field and is inverse proportional to the cubic distance r_{12}:

$$\hat{\mathcal{H}}_{dd} = g_e^2 \mu_B^2 \left\{ \frac{3(\boldsymbol{S}_1 \boldsymbol{r}_{12})(\boldsymbol{S}_2 \boldsymbol{r}_{12})}{r_{12}^5} - \frac{\boldsymbol{S}_1 \boldsymbol{S}_2}{r_{12}^3} \right\} = \boldsymbol{S}_1 \cdot \underline{\boldsymbol{D}} \cdot \boldsymbol{S}_2 \ . \tag{2.24}$$

$\underline{\boldsymbol{D}}$ denotes the dipole-dipole interaction tensor. Equation 2.24 represents the dipolar interaction between the electron magnetic moments in the point dipole approximation. The calculation of the spin operators in Equation 2.24 neglecting all non-secular terms leads to

$$\hat{\mathcal{H}}_{dd} = \omega_{dd} \boldsymbol{S}_z^1 \boldsymbol{S}_z^2 \quad \text{with}$$
$$\omega_{dd}(r_{12}, \theta) = \frac{\mu_B^2 g_e^2}{r_{12}^3} \left\{ 3 \cos^2(\theta) - 1 \right\} \tag{2.25}$$

for spherical coordinates with the already explained angle θ. Figure 2.6 illustrates the connection vector $\boldsymbol{r}_{12}$ and the enclosed angle θ of Equation 2.25.

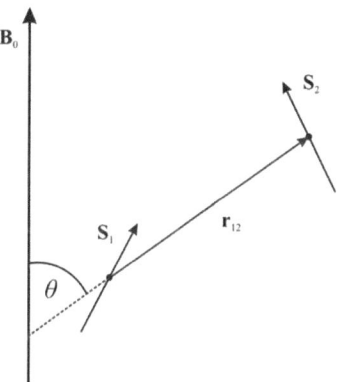

Figure 2.6.: Relative disposition of the magnetic moments $\boldsymbol{S}_1$ and $\boldsymbol{S}_2$. r_{12} is the distance between the two electron spins and θ the angle which is enclosed by $\boldsymbol{B}_0$ and $\boldsymbol{r}_{12}$.

The dipole-dipole tensor can be diagonalized in the main axial coordinate system and the resulting 3 × 3-matrix is then traceless:

$$\underline{D} = \frac{g_e^2 \mu_B^2}{r_{12}} \begin{pmatrix} -1 & & \\ & -1 & \\ & & 2 \end{pmatrix} = \begin{pmatrix} -\omega_{dd} & & \\ & -\omega_{dd} & \\ & & 2\omega_{dd} \end{pmatrix}. \qquad (2.26)$$

The dipole-dipole tensor can be used to measure electron-electron distances up to 8 nm[24] as the dipolar coupling frequency ω_{dd} is related to the electron spin-spin distance r_{12} (2.25). In liquids, the orientation dependence expressed by $\{3\cos^2(\theta) - 1\}$ averages out as the molecular rotation is fast. In solids, all angles between 0 and π emerge with the weighted probability $\sin(\theta)$. The characteristic spectrum resulting from the θ- and r_{12}-dependence is displayed in Figure 2.7 and is called Pake pattern.[25]

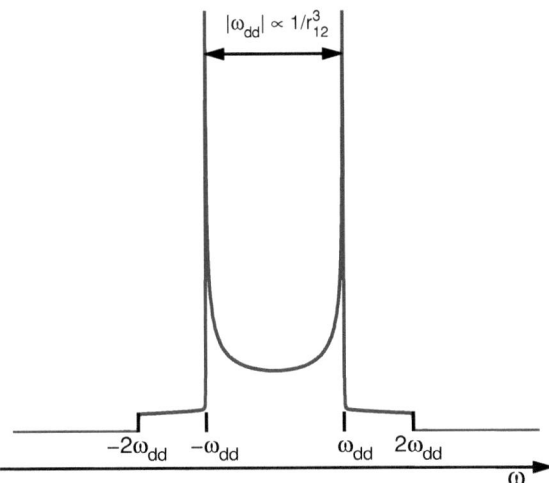

Figure 2.7.: Pake pattern as expected for a solid sample and an axial tensor $\underline{D}$. The singularities at $-\omega_{dd}$ and ω_{dd} are clearly visible.

The singularities at ω_{dd} can be used to determine directly the electron spin-electron spin distance via Equation 2.26.

The scalar coupling $\hat{\mathcal{H}}_J$ is another field-independent term. It is called *spin exchange* interaction or *Heisenberg spin exchange* (HSE) interaction. This kind of interaction mainly occurs for short distances ($r_{12} < 1.5$ nm). The interaction is caused by the Coulomb potential of the electron spins and is expressed by

$$\hat{\mathcal{H}}_J = J \boldsymbol{S}_1 \cdot \boldsymbol{S}_2 \ , \tag{2.27}$$

where J is a scalar term. J is the exchange integral of two spins $\boldsymbol{S}_1$ and $\boldsymbol{S}_2$

$$J = \left\langle \Phi_1(r_1)\Phi_2(r_2) \left| \frac{e^2}{r_{12}} \right| \Phi_1(r_2)\Phi_2(r_1) \right\rangle \ . \tag{2.28}$$

Equation 2.28 describes the mutual impact of the electron densities of spin 1 and 2 via Coulomb interaction.

The hyperfine interaction $\hat{\mathcal{H}}_{SI}$ arises as atomic nuclei also possess a magnetic moment μ_I as described in Section 2.1. The interaction between nuclear spins $\boldsymbol{I}$ and electron spins $\boldsymbol{S}$ is addressed as *hyperfine interaction* due to historical reasons. The *hyperfine interaction*

$$\hat{\mathcal{H}}_{SI} = \boldsymbol{S} \cdot \underline{\boldsymbol{A}} \cdot \boldsymbol{I} \ , \tag{2.29}$$

with the hyperfine tensor $\underline{\boldsymbol{A}}$, similar to the electron spin-electron spin interaction comprises an anisotropic (dipolar) part $\hat{\mathcal{H}}_{SI}^{dd}$ and an isotropic (scalar) part $\hat{\mathcal{H}}_{SI}^{iso}$. The anisotropic part is given by the analogue to Equation 2.24

$$\hat{\mathcal{H}}_{SI}^{aniso} = g_e \gamma_I \mu_B \mu_I \left\{ \frac{3(\boldsymbol{S}\boldsymbol{r}_{SI})(\boldsymbol{I}\boldsymbol{r}_{SI})}{r_{SI}^5} - \frac{\boldsymbol{SI}}{r_{SI}^3} \right\} \ , \tag{2.30}$$

where one electron spin was substituted by a nuclear spin $\boldsymbol{I}$. Equation 2.30 represents the interaction between the nuclear and electron magnetic moment in the point dipole approximation. The isotropic part

$$\hat{\mathcal{H}}_{SI}^{iso} = \hbar a_{iso} \boldsymbol{S} \cdot \boldsymbol{I} \qquad (2.31)$$

takes into account the effect that the electron is found at the nucleus of the atom, which is called the Fermi contact interaction. The Fermi contact interaction equals zero for a non-vanishing angular momentum.

The hyperfine interaction splits the energy levels in $2(I+1)$ sub-levels. In liquids, only the isotropic part remains due to the fast motion and rotation of the molecules, whereas, in solids, the isotropic as well as the anisotropic element can be measured. Therefore, the vicinity of the electron spins can be characterized by the hyperfine coupling.

The last term $\hat{\mathcal{H}}_{mw}$ denotes the manipulation of the total magnetization by the irradiation of microwaves (mw). The perturbing electromagnetic field $\boldsymbol{B}_{1e}$ propagates perpendicular to the static field $\boldsymbol{B}_0$. The microwave irradiation can be continuous which results in an excitation of the electron spins at appropriate frequency. This effect is utilized in continuous-wave (CW) EPR spectroscopy. On the other hand, the $\boldsymbol{B}_{1e}$ field can be chosen as a short pulse or a sequence of pulses where the strength and length of the $\boldsymbol{B}_{1e}$-pulse depend on the settings of the experiment. In EPR, the total magnetization vector can be rotated analogous to NMR by a *flip angle* $\beta = \omega_1 t_p$. The $\boldsymbol{B}_{1e}$ field for an angle $\pi/2$ and a duration of 16 ns at a microwave frequency of 9.7 GHz yields an amplitude of roughly 0.56 mT.

The EPR Spectrum of a Typical Nitroxide

The molecular structure of a typical nitroxide like TEMPOL (4-Hydroxy-2,2,6,6-tetramethyl-piperidine-1-oxyl) is shown in Figure 5.1. The main axes of the $\boldsymbol{g}$ tensor are drawn as well. The z-axis is aligned along the p-orbital at the ^{14}N atom, the x-axis along the NO bond and the y-axis is perpendicular to both.

The EPR spectrum of TEMPOL in water at room temperature (RT) is characterized by two terms:

$$E_{NO} = g\mu_B B_0 m_s + a_{iso} m_s m_I \; . \qquad (2.32)$$

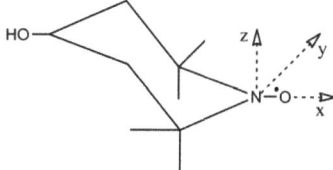

Figure 2.8.: Chair conformation of TEMPOL. The coordinate system of the main axes of the $\underline{g}$ tensor is depicted.

The isotropic hyperfine constant due to the coupling to the nitrogen nucleus is represented by a_{iso}, g is the isotropic g-factor and m_s and m_I are the magnetic spin-quantum numbers. m_s can only take the values $\pm 1/2$ whereas m_I depends on the number of neutrons in the nitrogen atom ($m_I = \pm 1/2$ for ^{15}N and $m_I = 0, \pm 1$ for ^{14}N). The CW EPR spectrum of ^{14}N-TEMPOL which exhibits three hyperfine lines is given in Figure 2.9. The hyperfine lines are equally separated by $a_{iso} = 15$ G due to motional averaging at room temperature.

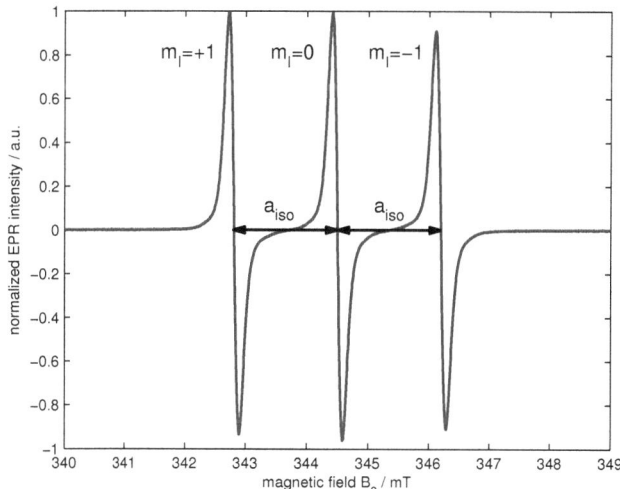

Figure 2.9.: CW EPR spectrum of ^{14}N TEMPOL in water at room temperature. The hyperfine couplings can be identified by the three resonance lines. The strong coupling a_{iso} leads to an obvious separation (~ 1.7 mT) of the hyperfine lines.

Pulsed EPR - DEER and ESE

Electron spin echo-detected (ESE) EPR and double electron-electron resonance (DEER) are well established techniques to investigate radical-radical distances and distance distributions in a sample containing paramagnetic agents. ESE and DEER measurements are recorded at cryogenic temperatures. In the frozen state the dipolar interactions do not average out and the distances between radicals can be assessed by relating the strength of the measured dipolar couplings to their separation ($\nu_{dd} \propto r^{-3}$). In ESE spectra dipolar interactions are not resolved but lead to an underlying broadening of the lines. For this reason only strong dipolar couplings can be extracted from linewidth analysis and ESE is only sensitive to distances up to 2 nm. Further, by the convolution of an unbroadened "reference" lineshape with a dipolar broadening function one can derive the dipolar coupling strength between the electron spins.

DEER is a four-pulse experiment in which the dipolar interactions are studied by measuring the electron spin echo at variable time delays between microwave pulses on another spin packet (double resonance). The dipolar couplings can be calculated from the DEER time domain data by Tikhonov regularization[24,26,27] or by direct simulation of the frequency spectra, both of which overcome the mathematically ill-posed problem of extracting distances and distributions. The dipolar coupling frequency is related to the electron spin pair distance via Equation 2.33

$$\nu_{dd}(r) = \frac{52.04 \text{ MHz} \cdot \text{nm}^3}{r^3}, \qquad (2.33)$$

where r is given in nm and reflects the electron-electron distance. Thus, DEER measurements are responsive to spin distances from 1.5 nm to roughly 8 nm. In this respect the EPR methods of choice can be seen as complementary and complete to investigate radical distance distributions in spin-labeled molecules in the range of 0 nm to 8 nm.[28] These results gained from the EPR experiments are then used to analyze and interpret the DNP experiments.

2.3. Dynamic Nuclear Polarization

NMR suffers severe inherent sensitivity problems due to the low magnetogyric ratio of nuclei. This can be overcome by several hyperpolarization techniques or by the *brute force* method. In this Subsection, the most common polarization methods and the theoretical background of DNP at low and at high temperatures will be given. Note that one has to differentiate between chemical and physical approaches for NMR signal enhancement.

The brute force method is the simplest but technically advanced way to generate a higher polarization. It simply relies on the Boltzmann distribution of the Zeeman energy levels. As the polarization increases linearly with the applied magnetic field strength and the inverse temperature as can be seen by Equation 2.9, the polarization can be multiplied by the factor of field enhancement or temperature reduction. This brute force approach has its limits as the technical realization of high magnetic fields and low temperatures is difficult. Further, the expenses for construction and maintenance are non-negligible. Up-to-date, the largest magnetic fields for NMR spectroscopy and clinical magnetic resonance imaging (MRI) are 24 T and 9 T, respectively. The lowest achievable temperatures are in the range of 1 K. This gives a polarization for protons of 2.45% and 80 ppm at 1 K and room temperature, respectively. Consequently, the possible signal enhancements at 1 K are still up to $\frac{1}{2.45\%} \approx 40$. For room temperature NMR measurements the possible signal enhancements can even exceed a factor of 10,000. For hetero nuclei and/or MRI, the possible enhancements are even higher. One has to keep in mind that for medical applications the cooling to cryogenic temperatures is not feasible as patients have to be investigated at physiological temperatures. Therefore, the possible signal enhancements for MRI in medicine are $> 30,000$ depending on the nucleus.

Para-hydrogen induced polarization (PHIP) is a chemical approach which transfers the spin order of para-hydrogen to unsaturated substrates by covalent binding. Theoretical signal enhancements of up to 10^5 are possible. In the early days of PHIP, the *para*-hydrogen had to bind to an unsaturated substrate which made it exclusively applicable to few molecules. Recently, Duckett *et al.*[5] established a method (SABRE) to overcome the need of unsaturated molecules for utilizing PHIP. Therefore, PHIP is almost as versatile as DNP but comprises non-trivial chemistry.

2. Theoretical Background

Chemically induced DNP (CIDNP) is based on a photo-chemical process in which transient radicals arise upon irradiating the sample with light. The CIDNP process is explained by the radical pair mechanism and is found in detail elsewhere.[29] It is mostly observed in liquids and the enhancements are approximately 10^2 fold. The limiting factors are the low enhancement factors and the necessity to produce radical pair intermediates by irradiation of light. Due to historical reasons, DNP and CIDNP share a common name but have nothing in common besides this.

Dynamic nuclear polarization (DNP) is a polarization technique which transfers the polarization of unpaired electron spins to the surrounding nuclei by microwave irradiation. The theoretical enhancement is given by the ratio of the magnetogyric ratios of the electron spin and a nuclear spin. This ratio is for proton (^{1}H), fluorine (^{19}F) and carbon (^{13}C) nuclei $E = 658 \approx 660$, $E \approx 700$ and $E = 2617 \approx 2600$, respectively. The most characteristic feature and significant drawback are the presence of unpaired electrons in the sample which have to be added if they are not inherent to the sample. On the other hand, DNP is not limited to any molecule or nucleus which makes it remarkably versatile. Several DNP techniques exist which can be divided into two subgroups:

1. *In situ* polarization methods, where polarization and NMR/MRI detection are implemented in the same magnet, namely solid-state DNP[30] and Overhauser DNP at high[31] and low magnetic field strength.[14,32]

2. *Ex situ* polarization methods, where polarization and NMR/MRI detection take place in different magnets, namely shuttle DNP[33] and dissolution DNP.[13,34]

Furthermore, one has to discriminate between the DNP polarization conditions, namely low-temperature or high-temperature DNP:

1. The polarization transfer at room temperature is mediated via the well-known *Overhauser effect*.[35–37]

2. The polarization at low-temperature is transferred through the *solid effect* (SE), *cross effect* (CE) or *thermal mixing* (TM).[12,38,39]

In the following Subsections, all DNP techniques will be introduced with a strong emphasis on the *Overhauser effect* as most DNP experiments of this work were performed at room temperature.

DNP at Room Temperature - Overhauser Effect

The *Overhauser effect* was first predicted by Albert Overhauser in 1953 for metals[35] and experimentally verified by Thomas Carver and Charles Slichter in 1956.[37] Abragam extended the theory of the *Overhauser effect* to non-metals in 1955.[36] The *Overhauser effect* in liquids with dissolved paramagnetic agents (unpaired electron spins = stable radicals) can be deduced from relaxation mechanisms in a simple two spin system with $S = I = 1/2$. For a better understanding of the *Overhauser effect*, the four energy levels of such a system are drawn schematically in Figure 2.10 with appropriate populations in equilibrium and its transition rates.

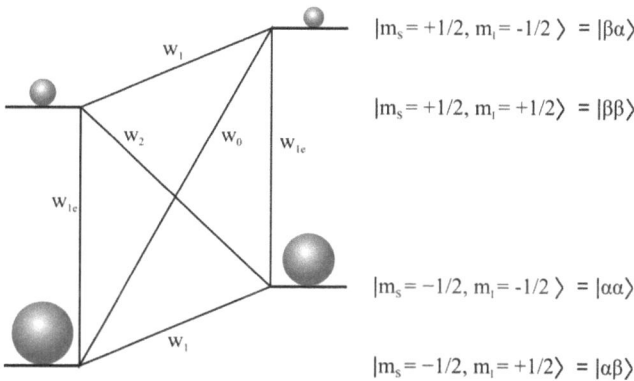

Figure 2.10.: The four energy levels of a two spin system. m_s and m_I denote the spin-quantum number of electron and nuclear spin, respectively. The levels are populated obeying the Boltzmann statistic. The possible relaxation paths are drawn with their appropriate transition rates. w_0 and w_2 are the zero- and double-quantum transition rates, respectively. w_1 and w_{1e} denote the single-quantum transition rates of a nuclear and an electron spin flip, respectively.

The transition rates of the coupled spin system can be calculated by utilizing the *Hamiltonian* of an interacting electron and nuclear spin in an external magnetic field:

$$\begin{aligned}
\hat{\mathcal{H}}_{OE} &= \hat{\mathcal{Z}}_S + \hat{\mathcal{Z}}_I + \hat{\mathcal{H}}_{SI} \\
&= \hbar g_S (\boldsymbol{S} \cdot \boldsymbol{B}_0) + \hbar \gamma_I (\boldsymbol{I} \cdot \boldsymbol{B}_0) \\
&\quad + \hbar^2 g_S^2 \gamma_I^2 \left\{ \frac{8\pi}{3} |\Psi(0)|^2 (\boldsymbol{S} \cdot \boldsymbol{I}) + \frac{3(\boldsymbol{S} \cdot \boldsymbol{r})(\boldsymbol{I} \cdot \boldsymbol{r})}{r^5} - \frac{\boldsymbol{S} \cdot \boldsymbol{I}}{r^3} \right\} \quad .
\end{aligned} \quad (2.34)$$

The first two terms are the electronic and nuclear Zeeman terms which determine the energy levels of the system as shown in Figure 2.10. They do not lead to relaxation phenomena and are subsequently ignored in the derivation of the *Overhauser effect*. The third term describes the scalar and dipolar interaction of the electronic and nuclear spin. It is the origin of relaxation phenomena and the DNP effect. Nuclear spin-spin interactions are neglected. The scalar part is proportional to the square of the wave function of the unpaired electron spin at the nucleus and connects the states $|\alpha\beta\rangle$ and $|\beta\alpha\rangle$. Namely, this term is responsible for the relaxation path w_0. Analogue, the dipolar part connects $|\alpha\alpha\rangle$ with $|\beta\beta\rangle$ and causes relaxation via the double-quantum transition w_2.

The transition rates introduced in Figure 2.10 can be utilized to calculate the total nuclear relaxation rate (reciprocal nuclear spin-lattice relaxation time)

$$1/T_1 = w^t = w_0 + 2w_1 + w_2 + w^0 \quad , \quad (2.35)$$

where w^0 denotes all relaxation paths which are not caused by the electronic spin $\boldsymbol{S}$. Furthermore, from Figure 2.10 the coupled differential Equations of the experimentally observables $\langle S_z \rangle$ and $\langle I_z \rangle$ are obtained

$$\begin{aligned}
\frac{d\langle I_z \rangle}{dt} &= -(w_0 + 2w_1 + w_2)(\langle I_z \rangle - I_0) - (w_2 - w_0)(\langle S_z \rangle - S_0) \\
\frac{d\langle S_z \rangle}{dt} &= -(w_2 - w_0)(\langle I_z \rangle - I_0) - (w_0 + 2w_1' + w_2)(\langle S_z \rangle - S_0) \quad ,
\end{aligned} \quad (2.36)$$

where S_0 and I_0 are the values of $\langle S_z \rangle$ and $\langle I_z \rangle$ in thermal equilibrium. The solution of

2. Theoretical Background

Equation 2.36 for the steady state $\frac{d\langle I_z \rangle}{dt} = 0$ was derived by Hausser and Stehlik:[40]

$$\langle I_z \rangle = I_0 \left\{ 1 + \frac{w_2 - w_0}{w_0 + 2w_1 + w_2} \cdot \frac{w_0 + 2w_1 + w_2}{w^t} \cdot \frac{S_0 - \langle S_z \rangle}{S_0} \cdot \frac{S_0}{I_0} \right\} . \tag{2.37}$$

It is customary to insert the following well-known abbreviations:

$$\begin{aligned} \text{Coupling parameter: } \xi &= \frac{w_2 - w_0}{w_0 + 2w_1 + w_2} \\ \text{Leakage factor: } f &= \frac{w_0 + 2w_1 + w_2}{w^t} \\ \text{Saturation factor: } s &= \frac{S_0 - \langle S_z \rangle}{S_0} . \end{aligned} \tag{2.38}$$

Furthermore, the ratio $\frac{S_0}{I_0}$ can be replaced by $\frac{\gamma_S}{\gamma_I}$. With these abbreviations we yield

$$\langle I_z \rangle = I_0 (1 + \xi f s \frac{\gamma_S}{\gamma_I}) \tag{2.39}$$

with the well-known enhancement factor E for the *Overhauser effect*

$$E = 1 + \xi f s \frac{\gamma_S}{\gamma_I} . \tag{2.40}$$

Equation 2.40 is the most important Equation for DNP at room temperature. Its components will be elaborated in the following paragraphs. Insertion of Equation 2.39 in the definition of the polarization leads to

$$P = \frac{I_0 - \langle I_z \rangle}{I_0} = \xi f s \frac{\gamma_S}{\gamma_I} , \tag{2.41}$$

where the maximum theoretical polarization enhancement is given by the quotient of the magnetogyric ratios $\frac{\gamma_S}{\gamma_I}$ for $\xi = f = s = 1$.

2. Theoretical Background

Leakage Factor

The leakage factor f is the part of the relaxation which is induced by the electronic spin. Therefore, it can be calculated by measuring the nuclear spin-lattice relaxation rate at the operating field with ($R_1 = w_1$) and without ($R_1^0 = w^0$) paramagnetic agent:

$$f = \frac{w_0 + 2w_1 + w_2}{w^t} = 1 - \frac{w^0}{w_1} = 1 - \frac{R_1^0}{R_1} \tag{2.42}$$

and $f(c \to \infty) = 1$ for high concentrations of the polarizing agent. As a result, the leakage factor f is not the limiting factor in DNP and it can be determined independently from T_{1n} inversion or saturation recovery experiments.

Saturation Parameter

The saturation parameter s gives the degree of equalization of the energy states due to the electronic Zeeman splitting. In Figure 2.10, this corresponds to the equalization of the states $|\alpha\beta\rangle$ and $|\beta\beta\rangle$ and $|\alpha\alpha\rangle$ and $|\beta\alpha\rangle$, respectively. One can safely assume that the transition frequencies of these states are similar and consequently can be saturated simultaneously. The saturation can be expressed in terms of the total magnetization and yields the steady-state solution of the familiar Bloch Equations:

$$s = \frac{S_0 - \langle S_z \rangle}{S_0} = \frac{M_0 - \langle M_z \rangle}{M_0} = \frac{\omega_{1e}^2 T_{1e} T_{2e}}{1 + \Omega_{0e}^2 T_{2e}^2 + \omega_{1e}^2 T_{1e} T_{2e}} \quad . \tag{2.43}$$

Here, $\Omega_{0e} = \omega_e - \omega_{0e}$ is the electronic offset frequency, T_{1e} and T_{2e} are the electron spin-lattice and spin-spin relaxation times and ω_{1e} is the amplitude of the transverse microwave field. This result is only valid for non-coupled EPR hyperfine lines which undergo no interaction. The maximum saturation results in the best enhancement. Ergo, the numerator of Equation 2.43 needs to be maximized and the denominator minimized. Minimization can be achieved by irradiating on-resonant ($\omega_e = \omega_{0e}$):

$$s = \frac{\omega_{1e}^2 T_{1e} T_{2e}}{1 + \omega_{1e}^2 T_{1e} T_{2e}} \quad . \tag{2.44}$$

Further, one can adjust s in the experiment by changing the used microwave power P_mw, namely $\omega_{1e} \propto P_\mathrm{mw}^{\frac{1}{2}}$. The extrapolation to infinite power will saturate the EPR transition completely:

$$s_{\max}(P_{\mathrm{mw}} \to \infty) = 1 \ . \tag{2.45}$$

This is ideal for a paramagnetic agent with only one hyperfine line and a microwave source with infinite power. However, in experiments the microwave power is limited due to technical reasons. Furthermore, the microwave power is restricted to moderate values as the electric component of the transverse electromagnetic field induces severe heating effects which finally leads to the boiling of the water sample. This well-known phenomenon (dielectric heating) is utilized in microwave ovens. Due to these effects, $s = 1$ cannot be realized in experiments. It is visualized in Figure 2.11 how s decreases from 0.93 to 0.34 when ω_{1e} is decreased by a factor of five. Furthermore, the red saturation profile is clearly power-broadened.

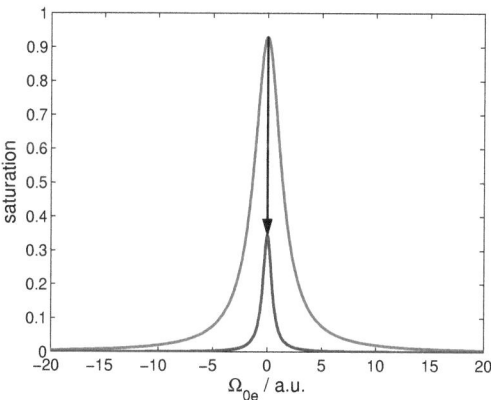

Figure 2.11.: Saturation profiles according to Equation 2.43 for a single non-coupled EPR hyperfine line. The blue saturation profile reflects the reduction of the achievable saturation on the decrease of ω_{1e} by a factor of five. The parameters of the calculation are: $T_{1e} = 1.7\,\mu s$, $T_{2e} = 0.4\,\mu s$ and $\omega_{1e} = 2\pi\gamma_e \cdot 0.05\,G$ and $\omega_{1e} = 2\pi\gamma_e \cdot 0.25\,G$, respectively.

The saturation parameter also depends on the electron spin-lattice and spin-spin relaxation times $s = s(T_{1e}, T_{2e})$. From Equation 2.44, one finds that increased relaxation times result in a higher saturation. Further, most used radicals are nitroxides as introduced in Figure 5.1. Nitroxides are characterized by three EPR hyperfine lines as seen in the CW EPR spectrum of the nitroxide TEMPOL (Figure 2.9). Thereupon, the saturation parameter for a single non-coupled hyperfine line modifies to

$$s = \frac{1}{3} \cdot \frac{\omega_{1e}^2 T_{1e} T_{2e}}{1 + \Omega_{0e}^2 T_{2e}^2 + \omega_{1e}^2 T_{1e} T_{2e}} \qquad (2.46)$$

to take account of three well-separated hyperfine lines. The saturation profile has the same shape as in Figure 2.11 only compressed by a factor of three. The achievable saturations become $s = 0.31$ and $s = 0.12$ for $B_{1e} = 0.25$ G and $B_{1e} = 0.05$ G, respectively. From a theoretical point of view for three non-coupled hyperfine lines only a saturation factor $s = \frac{1}{3}$ is achievable. Fortunately, there exist mainly two kind of interactions which couple the hyperfine lines. First of all, it is *Heisenberg spin exchange* (HSE) which mixes the lines if two radicals in solution collide:

$$A(\downarrow) + B(\uparrow) \longrightarrow A(\uparrow) + B(\downarrow) \ . \qquad (2.47)$$

Spin A and spin B exchange their spin state while the total spin number is preserved. The rate of the spin exchange is proportional to the concentration of radicals in the sample. HSE is effective because two spins change their polarization simultaneously. The other mechanism which mixes the states is electron spin-nuclear relaxation. In the case of nitroxides, the relaxation is caused by the coupling of the electron spin to the nitrogen nuclear spin. This coupling is also the origin for the hyperfine splitting as explained in Equation 2.32 and shown in Figure 2.9. As electron spin-nuclear relaxation flips only one spin, this mixing mechanism is not as effective as HSE. The electron spin-nuclear relaxation rate ($R_{1ne} = 1/T_{1ne}$) depends on the dynamics of the radical and consequently depends on the size of the radical itself and the viscosity of the solvent. Including the effects of HSE and electron spin-nuclear relaxation, the saturation of one hyperfine line can exceed the value of $s = \frac{1}{3}$ in the high microwave power limit as it was verified by Sezer *et al.*[31] The saturation s can be determined most accurately by performing pulse electron double resonance (ELDOR) experiments where one hyperfine line is irradiated (pumped) and another hyperfine line is observed. This allows for the quantitative calculation[41] of the effects of HSE and electron spin-nuclear relaxation. The last effect which can lead to a saturation exceeding $s = 1/3$ is a T_{2e} effect. Due to broad

EPR lines (short T_{2e}), the off-resonant hyperfine lines can contribute a non-negligible part to the total saturation in the limit of high microwave power. This phenomenon was observed for a spin-labeled macromolecule and will be discussed in more detail in Section 5.2.

Coupling Parameter

The coupling parameter ξ defines the coupling strength of the electron-nucleus interaction and is the most important quantity of DNP. As shown in Equation 2.38, ξ is strongly related to $w^t = 1/T_{1n}$. Consequently, it can in principle be determined by nuclear spin-lattice relaxation measurements.

Here, only the derivation of ξ for nitroxides in water will be given. One can safely assume the conditions of the *extreme-narrowing limit* and that the electron-nucleus interaction is of pure dipolar origin. The *extreme-narrowing limit* is valid for very short correlation times τ_c (very rapid molecular motion corresponds to $\omega\tau \ll 1$) where T_{1e} and T_{2e} become similar and the spectral density function $\mathcal{J}(\omega) = \frac{\tau_c}{1+\omega^2\tau_c^2}$ is (almost) constant. Under these conditions, it was shown by I. Solomon[42] that the transition rates have the following dependence:

$$\begin{aligned}
\text{Zero-quantum transition:} \quad & w_0 = const. \cdot 2 \cdot \mathcal{J}(\omega_S - \omega_I) \\
\text{Single-quantum transition:} \quad & w_1 = const. \cdot 3 \cdot \mathcal{J}(\omega_I) \\
\text{Double-quantum transition:} \quad & w_2 = const. \cdot 12 \cdot \mathcal{J}(\omega_S + \omega_I) \ ,
\end{aligned} \qquad (2.48)$$

where *const.* is a constant of no further interest. Inserting Equation 2.48 in 2.38 returns

$$\xi = \frac{12 \cdot \mathcal{J}(\omega_S + \omega_I) - 2 \cdot \mathcal{J}(\omega_S - \omega_I)}{2 \cdot \mathcal{J}(\omega_S - \omega_I) + 2 \cdot 3 \cdot \mathcal{J}(\omega_I) + 12 \cdot \mathcal{J}(\omega_S + \omega_I)} \ . \qquad (2.49)$$

Further, one can exploit the relations $\mathcal{J}(\omega_S + \omega_I) \approx \mathcal{J}(\omega_S - \omega_I) \approx \mathcal{J}(\omega_I) \approx \mathcal{J}(0)$ which gives the maximal coupling parameter for pure dipolar coupling in the *extreme-narrowing limit*:

$$\xi = \frac{12 - 2}{2 + 6 + 12} = \frac{1}{2} \ . \qquad (2.50)$$

Experimentally, the coupling factor can be determined by measuring the nuclear spin-lattice relaxation times with and without paramagnetic agent at the operating and at high magnetic field. The corresponding formula was derived by Hausser and Stehlik:[40]

$$\xi = \frac{5}{7}\left\{1 - \frac{2w_1}{R_1 - R_1^0}\right\} . \qquad (2.51)$$

ξ increases with increasing R_1 meaning that more effective relaxation agents yield a higher coupling parameter. $2w_1$ can be determined by measuring $R_1^{highfield} = 2w_1 + w^0$ and $R_1^{0,highfield} = w^0$ at the high field yielding $2w_1 = R_1^{highfield} - R_1^{0,highfield}$ and for the coupling parameter:

$$\xi = \frac{5}{7}\left\{1 - \frac{R_1^{highfield} - R_1^{0,highfield}}{R_1 - R_1^0}\right\} . \qquad (2.52)$$

The coupling factor depends on the operating field $\xi = \xi(\omega_S, \omega_I)$ as is obvious from Equation 2.49. Stepping to the high field $\omega_S \tau_c \ll 1$ does not hold any more while $\omega_I \tau_c \ll 1$ is still fulfilled which leads to a decrease of all spectral density functions containing ω_S. This results in a significant decrease of ξ as the denominator of Equation 2.49 contains $\mathcal{J}(\omega_I)$ which stays constant as long as the condition $\omega_I \tau_c \ll 1$ is satisfied. The magnetic field range at the steepest decrease of ξ lies between 0.1 and 1 T.

Once the coupling factor has been determined, the dynamics of the solvent and the radical can be extracted as the spectral density function also depends on the correlation time. The relation between translational motion τ, closest approach d of two unlike spins and the diffusion coefficients of the solvent molecules D_I and the radicals D_S is given by[43]

$$\tau = \frac{d^2}{D_I + D_S} \qquad (2.53)$$

utilizing the force-free model. The force-free model rationalizes the relation between ξ and the underlying molecular motions analytically.[43–45] It was used for the interpretation of DNP data with nitroxide radicals[33] and assumes that the interacting spins are situated at the centers of spherical spin-bearing molecules which undergo translational diffusion with respect to each other. Following the force-free model, the spectral density function can be written as

$$\mathcal{J}(\omega,\tau) = \frac{1 + \frac{5\sqrt{2}}{8}(\omega\tau)^{1/2} + \frac{\omega\tau}{4}}{1 + (2\omega\tau)^{1/2} + (2\omega\tau) + \frac{\sqrt{2}}{3}(\omega\tau)^{3/2} + \frac{16}{81}(\omega\tau)^2 + \frac{4\sqrt{2}}{81}(\omega\tau)^{5/2} + \frac{(\omega\tau)^3}{81}} \quad . \quad (2.54)$$

By fitting $\xi = \xi(\mathcal{J}(\omega,\tau))$ to DNP data, one can extract the appropriate correlation time τ. Consequently, if d can be estimated or determined from another experiment, local solvent diffusion coefficients can be determined with the DNP approach.

Three-Spin Effect

The DNP described so far arises from direct radical-nucleus coupling. However, in real systems not only nuclei of interest but also other nuclei are hyperpolarized in the manner described before. For example, the DNP effect on carbon is observed for a ^{13}C containing molecule dissolved in water. The (driven) interaction between these hyperpolarized protons and carbons can then provide positive enhancement to the carbon as protons and carbons possess a positive magnetogyric ratio. This is expressed in an Equation[46] derived from Equation 2.40

$$E_{I_2} = 1 + \xi_2^S f_2^S s \frac{\gamma_S}{\gamma_{I_2}} + \xi_2^1 f_2^1 \frac{\gamma_{I_1}}{\gamma_{I_2}}(1 - E_{I_1}) \quad , \quad (2.55)$$

where S denotes the electronic spin, I_2 the detected nuclear spin and I_1 the third spin which is responsible for the three-spin effect. In Equation 2.55, the second summand reflects the well-known direct DNP term whereas the third term is the origin of the three-spin effect. The three-spin or indirect DNP term depends on the interaction strength between the unlike nuclear spins expressed by f_2^1 and ξ_2^1 and on the degree of hyperpolarization expressed by E_{I_1}. A simplified form of Equation 2.55 expressed in terms of the DNP and three-spin term has the form $E_{I_2} = 1 + \text{DNP term} + \text{three-spin term}$.[47] The absolute enhancement E_{I_2} in a three-spin system depends on the signs of the DNP and the three-spin term. Equal signs result in an increased total enhancement whereas different signs lower the total enhancement. As the signs of two terms is based on the type of the coupling (scalar or dipolar) and on the magnetogyric ratio, one has to discriminate between different three-spin systems. Here, the example mentioned above is reconsidered ($I_1 = {}^1$H, $I_2 = {}^{13}$C). The magnetogyric ratios of ^{1}H and ^{13}C are both positive.

Therefore, the direct DNP term is negative as one observes for nitroxides and molecules dissolved in water a negative enhancement for carbon. For protons, E_{I_1} is negative and $\frac{\gamma_{I_1}}{\gamma_{I_2}}$ is positive leading to a positive sign. Under these conditions, the indirect DNP term depends on the kind of coupling between the proton and the carbon spins. According to Brunner et al.,[46] the inter-nuclear coupling can be safely assumed to be $\xi_2^1 = +1/2$ for fast molecular motion.[48]

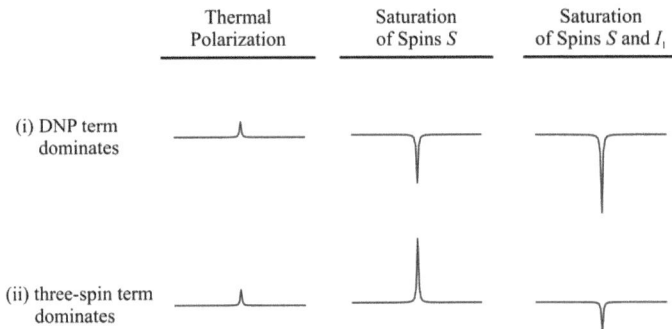

Figure 2.12.: A schematic to show the signs and magnitudes of the observed NMR signal at thermal equilibrium (without saturation), with saturation of spins S and with saturation of spins S and spins I_1. (i) If the DNP term dominates, the signal upon microwave irradiation is inverted and enhanced. After a double saturation, the NMR signal is even further enhanced. (ii) If the three-spin term dominates, the DNP-enhanced and equilibrium NMR signal will have the same sign. Upon saturation of spins S and I_1, the resulting signal will give a decreased and inverted sign.

The strength of the three-spin effect depends on the radical concentration. For high radical concentrations, the leakage factor f_2^1 becomes negligible as compared to f_2^S. Consequently, for high radical concentration $f_2^1(c \to \infty) \longrightarrow 0$ is valid and the three-spin term can be neglected. In summary, for the investigated system the total enhancement comprises a negative value of the direct DNP term and a positive sign of the three-spin term:

$$E_{I_2} = 1 - |\text{DNP term}| + |\text{three-spin term}| \quad . \tag{2.56}$$

Accordingly, a dominant DNP term leads to an inversion of the enhanced signal and a dominant three-spin effect gives a positive enhanced NMR signal. It should be mentioned that the three-spin term can be effectively removed by the saturation of the I_1 spins. The complete reasoning is visualized in Figure 2.12. For the experiments involving a saturation of the S and I_1 spins a triple-resonant probehead is required.

Solid-State DNP

At cryogenic temperatures, most samples are solids and the polarization transfer mechanism differs from the *Overhauser effect* as the paramagnetic centers are spatially fixed. DNP in solid dielectrics comprises the polarization transfer mechanisms *solid effect* (SE), *cross effect* (CE) and *thermal mixing* (TM) which will be introduced by following Duijvestijn et al.[49] One should note the following features of the DNP processes:

- *Solid effect:* Two-spin process involving forbidden flip-flop transitions.
- *Cross effect:* Three-spin process involving exclusively allowed transitions.
- *Thermal mixing:* Three-spin process involving forbidden flip-flop transitions.

The kind of mechanism which is responsible for the DNP effect in solids depends on the EPR and NMR properties of the sample. Namely, it depends on the homogeneous line width δ and the inhomogeneous spectral breadth Δ of the polarizing agent and on the nuclear Larmor frequency ω_{0I} of the nucleus which is polarized. The relation of these three parameters provides information which of the DNP mechanisms dominates:

$$
\begin{aligned}
&\text{\textit{Solid effect:}} \quad && \delta, \Delta < \omega_{0I} \\
&\text{\textit{Cross effect:}} \quad && \delta < \omega_{0I} < \Delta \\
&\text{\textit{Thermal mixing:}} \quad && \delta > \omega_{0I} \ .
\end{aligned}
\quad (2.57)
$$

For $\delta \approx \omega_{0I}$, a mixing of the mechanisms occurs. The DNP experiments in this work were performed at low magnetic field which results in a low nuclear Larmor frequency ω_{0I}. Therefore, the condition $\delta < \omega_{0I}$ is not fulfilled and the dominant effect is *thermal mixing*. Consequently, a short derivation for *thermal mixing* is given. Detailed derivations of the *solid effect*[38,50,51] and the *cross effect*[52,53] are given elsewhere.

Thermal Mixing

In the *thermal mixing* process, the electron polarization is not transferred directly to the nuclei but is mediated indirectly via dipole-dipole interactions of electron spins. With decreasing temperature and increasing radical concentration, the EPR line is broadened. For this short derivation a homogeneous line broadening is assumed but it is not a necessary condition.

The electron polarization is transferred by a two-step process which involves two interacting electronic spins and one nuclear spin. In a first step, the spin temperature of the dipolar electron bath is reduced by the irradiation of microwaves. In a second step, the spin temperatures of the dipolar electron bath and the nuclear Zeeman reservoir equalize which corresponds to the actual polarization enhancement. The introduction of coupled energy reservoirs is best described by the concept of spin temperature theory.[12,38] It is introduced via the density matrix

$$\sigma_{eq} = \exp(-\beta\mathcal{H})/Tr(-\beta\mathcal{H}) \quad , \tag{2.58}$$

which describes the spin system in equilibrium and is defined by

$$\beta = \frac{\hbar}{k_B T_S} \quad . \tag{2.59}$$

T_S is the spin temperature of the energy reservoir and the so-called inverse temperature β describes the degree of polarization of the spin system. Low spin temperatures correspond to high polarizations. The pivotal Equations for the explanation of *thermal mixing* are the Provotorov Equations plus a relaxation term

$$\dot{p}_e = -W\left(p_e - \frac{1}{2}\beta\Delta\right) - W_e\left(p_e - P_e^0\right) \quad \text{and} \tag{2.60}$$

$$\dot{\beta}_e = W\frac{2\Delta}{\omega_D^2}\left(p_e - \frac{1}{2}\beta\Delta\right) - W_{1D}\left(\beta - \beta^0\right) \quad , \tag{2.61}$$

which describe the time evolution of the inverse temperatures p_e and β. All parameters and their meanings are summarized in Table 2.1 and visualized in Figure 2.13.

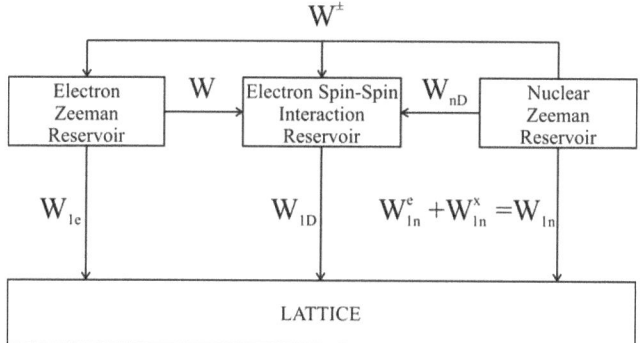

Figure 2.13.: Interactions between nuclear Zeeman, electron Zeeman and electron spin-spin interaction reservoirs.

The stationary solutions of these Equations are

$$p_e^\infty = p_e^0 \frac{W_{1e} + W\left(\frac{\Delta^2}{a\omega_D^2} - \frac{\Delta}{\omega_e}\right)}{W_{1e} + W\left(\frac{\Delta^2}{a\omega_D^2} + 1\right)} \quad \text{and} \tag{2.62}$$

$$\beta^\infty = -\frac{2p_e^0}{\omega_e} \frac{W_{1e} - W\frac{\omega_e}{\Delta}\left(\frac{\Delta^2}{a\omega_D^2} - \frac{\Delta}{\omega_e}\right)}{W_{1e} + W\left(\frac{\Delta^2}{a\omega_D^2} + 1\right)}. \tag{2.63}$$

With the requirement that $W_{1D} \gg C_n \omega_n^2/\omega_D^2$ (C_n = number of nuclei per electron), the differential Equation for the nuclear spin polarization p_n can be written as

$$\begin{aligned}\dot{p}_n = &- W^+\left(p_n - p_e - \frac{1}{2}\beta\left(\omega_n - \Delta\right)\right) - W^-\left(p_n + p_e - \frac{1}{2}\beta\left(\omega_n + \Delta\right)\right) \\ &- W_{nD}\left(p_n - \frac{1}{2}\beta\omega_n\right) - \left(W_{1n}^e + W_{1n}^x\right)\left(p_n - P_n^0\right) \quad ,\end{aligned} \tag{2.64}$$

where the nuclear spin-lattice relaxation rate $W_{1n} = W_{1n}^e + W_{1n}^x$ was divided into a part

Table 2.1.: Involved parameters of the stationary solution of the Provotorov Equations.

p_e	electron spin polarization
β	inverse spin temperature of the dipolar electron spin system
W	transition rate between the electron Zeeman and electron spin-spin interaction reservoir
W_{1e}	electron spin-lattice relaxation rate
W_{1n}	$= W_{1n}^e + W_{1n}^x$ nuclear spin-lattice relaxation rate
W_{1n}^e	nuclear spin-lattice relaxation rate induced by fixed paramagnetic centers
W_{1n}^x	nuclear spin-lattice relaxation rate induced by all other mechanisms besides the electron spins
W_{1D}	longitudinal relaxation rate of the electron spin-spin interaction reservoir
W_{nD}	transition rate between the nuclear Zeeman and electron spin-spin interaction reservoir
$W^{\pm}$	rates of the zero-quantum and double quantum transitions, respectively
a	$= W_{1D}/W_{1e}$
ω	microwave irradiation frequency
ω_e	electron Larmor frequency
ω_D	frequency corresponding to the strength of the electron spin-spin interaction reservoir
Δ	$= \omega - \omega_e$, microwave offset frequency
N_e	number of electron spins per volume unit

induced by the electron spins W_{1n}^e and a second part W_{1n}^x caused by all other mechanisms. The stationary solution for the nuclear polarization which describes *thermal mixing* as well as the *solid effect* has the form

$$p_n^\infty = p_n^0 \left(1 + \alpha \cdot p_n^1\right) \quad with \tag{2.65}$$

$$p_n^1 = \frac{\gamma_e}{\gamma_n} \cdot \frac{(W^+ - W^-)W_{1e} + (W^+ + W^- + W_{nD}) \cdot W \cdot \frac{\omega_n \Delta}{a\omega_D^2}}{(W^+ + W^- + W_{nD} + W_{1n}^e) \cdot \left(W_{1e} + W \cdot \left(\frac{\Delta^2}{a\omega_D^2} + 1\right)\right)} \quad and \tag{2.66}$$

$$\alpha = \frac{W^+ + W^- + W_{nD} + W_{1n}^e}{W^+ + W^- + W_{nD} + W_{1n}^e + W_{1n}^x} \;. \tag{2.67}$$

The first part $(W^+ - W^-) \cdot W_{1e}$ of Equation 2.66 is responsible for the *solid effect* and can consequently be neglected. The second term describes the effect of *thermal mixing* where W^+ and W^- compared to W_{nD} can be neglected for low microwave power. For high radical concentrations, $W_{1n}^x \ll W_{1n}^e, W_{nD}$ is valid and α becomes close to unity. With these approximations, the possible enhancement of the *thermal mixing* process yields

$$\begin{aligned} E &= \frac{p_n^\infty}{p_n^0} \\ &= 1 + \frac{\gamma_e}{\gamma_n} \cdot W_{nD} \cdot T_{1n} \cdot \frac{WT_{1e}\frac{\omega_n \Delta}{a\omega_D^2}}{1 + WT_{1e}\left(1 + \frac{\Delta^2}{a\omega_D^2}\right)} \;, \end{aligned} \tag{2.68}$$

where $T_{1n} = (W^+ + W^- + W_{nD} + W_{1n}^e)^{-1}$ and T_{1e} denote the nuclear and electron spin-lattice relaxation times, respectively. The nuclear spin polarization will have the same value throughout the sample if the spin diffusion within the solvent is very fast. This is the case for the samples investigated in this thesis. In the following paragraph, the meaning of the parameters is described and some easy-to-use formulas are given.

Thermal mixing is based on a good thermal contact between the energy reservoirs given in Figure 2.13. For $\delta \gg \omega_n$, a good thermal contact between the nuclear Zeeman and the electron spin-spin interaction bath can be assumed even in the absence of microwave irradiation as the homogeneous EPR line contains the nuclear Larmor frequency. An equalization between the electron Zeeman and the electron spin-spin interaction polarization can be optimized by matching the microwave frequency ω to the maximum of the EPR line differing by the electron spin-spin interaction frequency ω_D:

$$\Delta = \omega - \omega_e = \pm \omega_D \ . \tag{2.69}$$

Matching the frequency $\omega = \omega_e \pm \omega_D$ results in the highest NMR signal enhancement. The microwave frequency matching the electron Larmor frequency $\omega = \omega_e$ yields zero enhancement. Therefore, ω_D can be obtained from a plot of $E(\omega)$ and $E(B_0)$, respectively. This procedure is further explained in Chapter 6.

According to Duijvestijn et al.,[49] the factor $a = \frac{T_{1e}}{T_{1D}}$ is between one and three. WT_{1e} can be obtained by measuring the enhancements in dependence of the microwave power P. Subsequently, a plot of $(1-E)^{-1}$ versus the inverse power P^{-1} yields a straight line as

$$(1-E)^{-1} = const. \cdot \left\{ \left(1 + \frac{\Delta^2}{a\omega_D^2}\right) + \frac{1}{WT_{1e}} \right\} \text{ and} \tag{2.70}$$

$$W = \pi g(\Delta) \omega_{1e}^2 = \pi g(\Delta) \gamma_e^2 B_{1e}^2 = \pi g(\Delta) \gamma_e^2 \kappa P \ . \tag{2.71}$$

Here, $g(\Delta)$ denotes the normalized EPR line, B_{1e} the amplitude of the transverse microwave field and the proportionality factor κ gives the efficiency of the EPR probehead. The maximum achievable enhancement occurs at infinite power. The intercept of the straight line of Equation 2.70 with the y-axis yields the maximum achievable enhancement at full saturation (infinite power). W is subsequently determined by using

$$\frac{E}{E_{max}} = \left[1 + \frac{1}{WT_{1e}\left(1 + \frac{\Delta^2}{a\omega_D^2}\right)} \right]^{-1}, \tag{2.72}$$

where T_{1e} is determined independently from inversion recovery measurements.
It is difficult to calculate W_{nD} directly via

$$W_{nD} = \frac{3}{10}\omega_n^{-2}\gamma_e^2\gamma_n^2\hbar^2 \frac{1}{b^3 R^3} \frac{1}{T_{2e}^{SS}} \frac{\int_{-\infty}^{\infty} g(\omega)g(\omega-\omega_n)d\omega}{g(0)} \quad \text{with} \qquad (2.73)$$

$$T_{2e}^{SS} = \left(3.8\gamma_e^2\hbar N_e\right)^{-1},$$

$$R^3 = \left(\frac{4}{3}\pi N_e\right)^{-1} \quad \text{and}$$

$$b^3 = \frac{3}{10}\frac{\gamma_e^2\hbar^2}{B_0^2}\frac{1}{4\left\langle|q|^2\right\rangle R^3}.$$

$4\left\langle|q|^2\right\rangle$ can be calculated by comparison of

$$4\left\langle|q|^2\right\rangle = \frac{W^\pm}{\frac{1}{2}W(g(\Delta\pm\omega_n)/g(\Delta))} \quad \text{and} \qquad (2.74)$$

$$W^\pm = T_{1n}^* - T_{1n}. \qquad (2.75)$$

Here, T_{1n}^* is the nuclear spin-lattice relaxation time under the influence of microwave irradiation at frequency $\omega = \Delta\pm\omega_n$. All other parameters in Equation 2.68 are known. The transition rate W_{nD} can be more easily determined experimentally from Equation 2.68 if all values are known. Thus, W_{nD} was calculated choosing the latter method.

3. Technical Aspects of the Mobile Set-up

One major issue of the usage of hyperpolarized substances, especially in medical imaging, is the limited lifetime of the hyperpolarized state restricting the *in vivo* application and detection of the hyperpolarized molecules to roughly three times T_{1n}. This problem is even more pronounced if the hyperpolarization process cannot take place in the vicinity of the used MR scanner due to safety restrictions or space limitation like in medical MR facilities. In this thesis, the development of a mobile DNP polarizer based on an intermediate-field Halbach magnet[54] is reported. The use of a mobile permanent magnet design enables the minimization of the transport time from the hyperpolarization site to the object of interest (*e.g.* animal) resulting in an efficient use of the non-equilibrium magnetization. Due its mobility, the DNP polarizer can be removed from the MR scanner if the MRI system is used for routine scanning. In this Chapter, the basic set-up of a low-cost apparatus and its components for the experimental work are given. The first Section gives an overview of the complete set-up and some experimental details. The Halbach and an electromagnet are compared in Section 3.2, while the last Sections show imperfections and possible improvements of the set-up.

3.1. Experimental Set-up

The set-up of a (mobile) DNP polarizer comprises in general at least one magnet, a double resonant probehead for EPR transmission and NMR detection, a microwave generator and an NMR spectrometer. The number of used magnets depends on the method of DNP, whether the polarization takes place in the same[14,39] or in a different NMR detection magnet.[55,56] Here, all experiments were exclusively performed in one magnet which means that the hyperpolarization of the sample and the NMR detection took place in the same magnet.

In this work, for the EPR irradiation and NMR/EPR detection a Bruker (Karlsruhe, Germany) probehead (MD4EN), originally designed for electron-nuclear double resonance (ENDOR), was used. This probehead was used exclusively for DNP unless described otherwise and will be introduced in more detail in Section 3.4. For the DNP and EPR experiments, the field strength of a Bruker electromagnet or of the Halbach magnet (following Subsection) was adjusted to 0.345 T, corresponding to electron and ^{1}H Larmor frequencies of 9.7 GHz and 14.7 MHz, respectively. For NMR detection, a low field spectrometer (Kea, Magritek, Wellington, New Zealand) was employed (1 − 100 MHz). Continuous microwave irradiation in a critically coupled probehead was used to perform the DNP experiments in the described set-up. A Hewlett Packard HP 8350 B microwave source capable of a maximum output power of 30 mW and with a tuning range of 1 − 30 GHz was used. The source output was amplified by a Varian traveling wave tube (TWT) amplifier with a maximum continuous wave output power of 20 W at X-band frequencies (8 − 12.4 GHz). The mobile set-up is depicted in Figure 3.1 where it is placed in front of a tomograph at the University Clinics of Mainz.

3.2. Halbach Magnet versus Electromagnet

A special arrangement of permanent magnets, which results in an enhancement of the magnetic field on one side and an annihilation on the other side, is called a Halbach arrangement. The simplest superposition of the magnetic field of five permanent magnets is shown in Figure 3.2.

This arrangement is the starting point for more sophisticated arrangements. The assembly of permanent magnets in a ring design can achieve almost any configuration of the superimposed magnetic field, *e.g.* a dipolar or quadrupolar field profile within the ring. Figure 3.3 shows the arrangement generating a dipole field within the ring and without a significant stray field.

For the development of a mobile DNP polarizer, a dipolar Halbach magnet was designed out of two movable, stacked rings with sintered cobalt-samarium alloy permanent magnets as it is drawn in Figure 3.4. The remanence of a single permanent magnet is approximately 1.42 T. The simulation for the positioning of the single rings and each single magnet was performed by Dr. Peter Blümler (University of Mainz) in order to obtain a homogeneous total magnetic field.

The total magnetic field vector stands perpendicular to the bore of the cylindric Halbach magnet which is aligned along the z-axis. The theoretical magnetic field along the bore

3. Technical Aspects of the Mobile Set-up

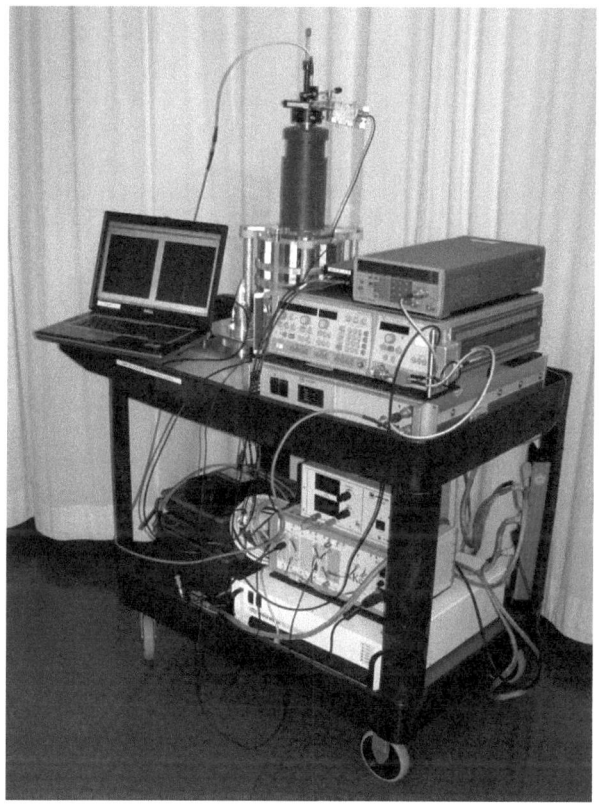

Figure 3.1.: Mobile DNP polarizer. The complete polarizer is mounted on one rack - the tunable Halbach magnet, the microwave source, the amplifier and the probehead can be seen on the upper tray. The NMR spectrometer is placed on the lower tray. The power supply for the LabVIEW-controlled switches is on top of the NMR spectrometer.

3. Technical Aspects of the Mobile Set-up

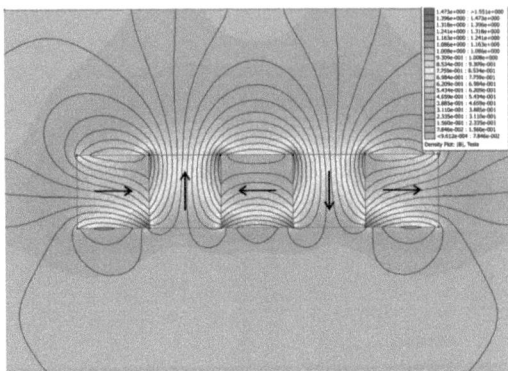

Figure 3.2.: Simplest Halbach arrangement. The arrows give the direction of the magnetic field of the permanent magnets in the point dipole approximation. Blue area corresponds to a low magnetic field and red area to a high magnetic field. It is obvious that the magnetic field on the top (red and yellow area) is enhanced whereas it cancels out at the bottom (blue and green area). This is achieved by turning neighboring magnets by $\pi/2$.

as calculated with a home-written *Mathematica* program is shown in Figure 3.5 and predicts a high homogeneity at the center in an interval of ± 30 mm.

The magnetic field strength of the Halbach magnet can be varied from 0.15 up to 0.46 T by moving the outer ring. One should note that the direction of the resulting field will change if the outer ring is moved. The physical properties of the mobile Halbach magnet are listed in Table 3.1 and are compared to a commercially available electromagnet. The advantages of the Halbach magnet are its compactness, its relatively low weight as well as its mobility as an external power supply and a cooling unit are not necessary. Furthermore, it is considerably cheaper than an electromagnet. The only two drawbacks of Halbach magnets are their - in comparison to an electromagnet - inhomogeneity and the dependence of its magnetic field strength on the ambient temperature. The Halbach magnet showed a drift of up to 15 G within 12 hours of measurement time (data not shown). The homogeneity of the used Halbach magnet and the electromagnet is compared quantitatively by line width analysis of an NMR spectrum as presented in Figure 3.6. The fitting results manifest that the line width in the electromagnet is approximately five times better than in the Halbach magnet. The homogeneity of the Halbach magnet can be improved using an active or passive shim unit. This topic is

3. Technical Aspects of the Mobile Set-up

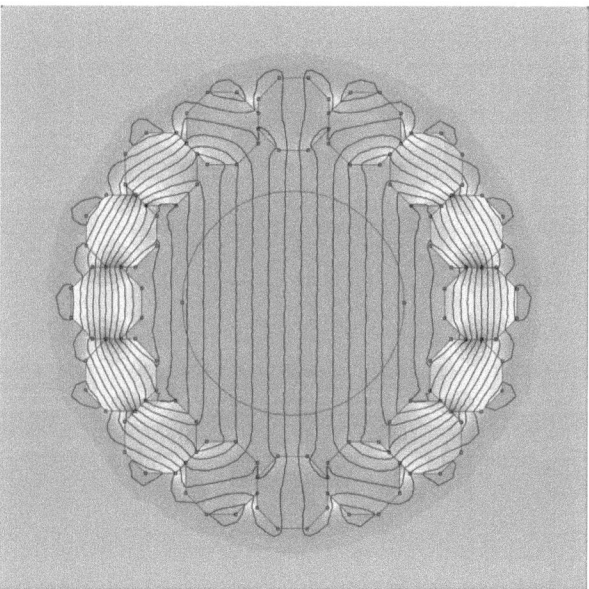

Figure 3.3.: Magnetic field of a ring of permanent magnets. The interior possesses an almost homogeneous dipole field whereas the stray field is nearly zero. The inner circle denotes the bore of the magnet ring.

3. Technical Aspects of the Mobile Set-up

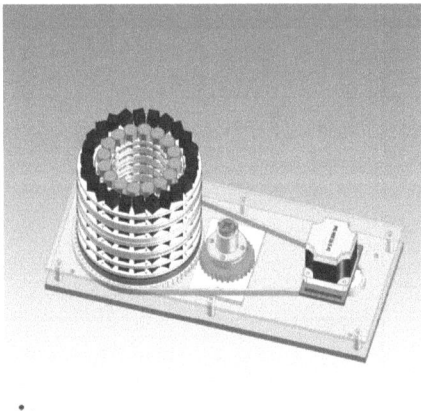

Figure 3.4.: CAD scheme of the mobile Halbach magnet including the step motor to move the outer ring.

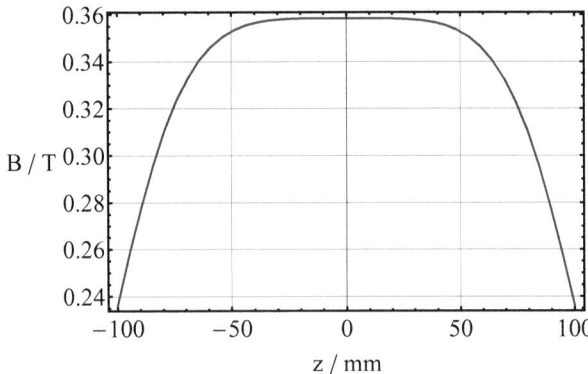

Figure 3.5.: Theoretical magnetic field of the Halbach magnet along the bore (z-axis). Obviously, the magnetic field at the center shows the highest homogeneity.

3. Technical Aspects of the Mobile Set-up

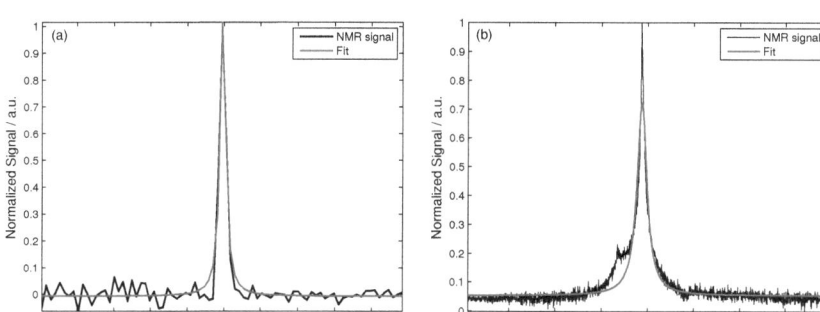

Figure 3.6.: Proton NMR signal in the electromagnet (left) and the Halbach magnet (right). (a) Fit of a Voigtian line shape (red line) to the NMR signal in an electromagnet yielding a full-width at half-maximum of less than 0.1 kHz. (b) The NMR spectrum in the Halbach magnet was fitted with a pseudo-Voigtian function (red line) to obtain a better fit. This fit yielded a full-width at half-maximum of less than 0.65 kHz.

Table 3.1.: Physical properties of the used Halbach magnet and a commercially available electromagnet.

Facts	Halbach	Electromagnet
Magnetic Field Strength [T]	$0.15 - 0.46$	$0 - 0.6$
Homogeneity [ppm]	~ 33	~ 7
Dimensions [mm]	$270 \times 320 \times 600$	$700 \times 800 \times 1300$
Weight [kg]	47	2100
Cooling Unit	not necessary	water
Costs [€]	10,000	100,000

3.3. Shimming of a Halbach Magnet

This Section briefly summarizes how a Halbach magnet can be shimmed by utilizing permanent magnets and spherical harmonics. Other possibilities to shim a magnet are additional coils and μ−metal which is a passive shimming method. The homogeneity of a magnet is essential for a sufficient signal-to-noise ratio when it comes to the detection of low γ nuclei like *e.g.* ^{13}C.

Active Shimming with Permanent Magnets

The term *shimming* comprises the improvement of the homogeneity of a magnetic field by all techniques and methods. In modern NMR magnets, the homogeneity is achieved with shim coils that can generate dipolar, quadrupolar and octopolar fields. In a mobile magnet, the current for the coils is obviously not available. Therefore, active shimming can be performed with additional permanent magnets which are also aligned in a Halbach arrangement in which each magnet can be moved individually in all three dimensions as it is demonstrated by Danieli et al.[57] Thereby, the inhomogeneities can be compensated by the superposition of the extra magnetic field. The exact shim procedure is described in the following:

1. Measurement of the magnetic field along the bore designated as z-axis and on a sphere.

2. Decomposition of the measured magnetic field into its multipole terms (spherical harmonics) which subsequently yields the coefficient of each term.

3. Optimization of the positioning of the magnets in the shim unit in order to compensate for the inhomogeneities.

4. Repeated iteration of the previous steps until the needed homogeneity is achieved.

The decomposition of the magnetic field was done in *Mathematica 6* with a custom-written program. The optimization of the positioning was performed utilizing the software package Radia which allows for the calculation and superposition of magnetic fields of permanent magnets. The used approach follows the one suggested by Danieli *et al.*[57] in which the correlation between the spherical harmonics and their Cartesian spatial dependence is introduced up to the second order:

$$\begin{aligned} B_0(\boldsymbol{r}) &= B_{00} + \Delta B_0(\boldsymbol{r}) \\ &= C_0 + C_{11+}x + C_{11-}y + C_{10}z + C_{20}(2z^2 - x^2 - y^2) \\ &\quad + C_{22+}(x^2 - y^2) + C_{22-}xy + C_{21+}xz + C_{21-}yz + o(r^3) \ . \end{aligned} \quad (3.1)$$

3. Technical Aspects of the Mobile Set-up

Here, $B_0(\boldsymbol{r})$ is a scalar as the other two components are negligible. B_{00} is the mean static field and $\Delta B_0(\boldsymbol{r})$ describes the inhomogeneity. The C's are the coefficients of the spherical harmonics and the spherical harmonics themselves are expressed in Cartesian coordinates. The effect of the shim unit which consists of eight permanent magnets by shifting magnet blocks is summarized in Table 3.2.

Table 3.2.: Shim coefficients versus spatial movements of the individual permanent magnets.

Coefficient	Spherical harmonics (Cartesian coordinates)	Magnet label							
		1A	2A	3A	4A	1B	2B	3B	4B
C_{11+}	x	$-\Delta x$	$-\Delta x$	$-\Delta x$	$-\Delta x$	Δx	Δx	Δx	Δx
C_{11-}	y	$-\Delta y$	$-\Delta y$	$-\Delta y$	$-\Delta y$	Δy	Δy	Δy	Δy
C_{10}	z	Δz	Δz	Δz	Δz	Δz	Δz	Δz	Δz
C_{21+}	xz	$-\Delta x$	$-\Delta x$	Δx	Δx	$-\Delta x$	$-\Delta x$	Δx	Δx
C_{21-}	yz	$-\Delta y$	$-\Delta y$	Δy	Δy	$-\Delta y$	$-\Delta y$	Δy	Δy
C_{22-}	xy	$-\Delta x$	Δx	$-\Delta x$	Δx	$-\Delta x$	Δx	$-\Delta x$	Δx
C_{22+}	$x^2 - y^2$	Δz	Δz	$-\Delta z$	$-\Delta z$	$-\Delta x$	Δx	$-\Delta x$	Δx
C_{20}	$z^2 - x^2 - y^2$	Δz	Δz	$-\Delta z$	$-\Delta z$	Δx	$-\Delta x$	Δx	$-\Delta x$

Technical Advances and Outlook

The shim unit which consists of eight permanent magnets was designed and constructed by Hans-Peter Raich. To this date, the construction has not been finished therefore no results can be shown. A picture of the first design and its CAD scheme are shown in Figure 3.7. The shim unit was originally planned for shimming a previous Halbach magnet which suffered from a high degree of inhomogeneity. Figure 3.8 illustrates the difference of the homogeneity of the two Halbach magnets. The old Halbach magnet shows a homogeneity of ~ 660 ppm whereas the new Halbach magnet shows an improved homogeneity of ~ 33 ppm although they are identical in design and used materials. Originally, the shim unit was planned to improve the homogeneity of the old magnet below 100 ppm. Due to the fact that the new Halbach magnet has an inherent homogeneity of ~ 33 ppm, the project of building a shim unit became obsolete during the thesis. The improved Halbach magnet is the first and most essential technical improvement achieved during the thesis. Without a homogeneous magnetic field, large DNP enhancements could not be observed and the detection of hetero nuclei were not even feasible. Therefore, the shim unit is only mentioned for completeness and to thus keep in mind the possibility

3. Technical Aspects of the Mobile Set-up

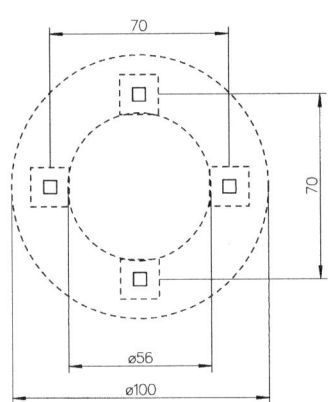 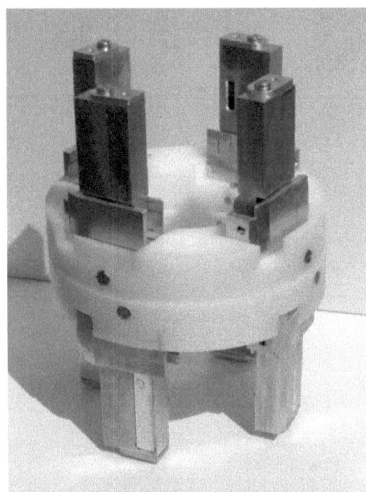

Figure 3.7.: Left: Top view of the shim unit showing four permanent magnets (magnet volume: $5 \times 5 \times 20$ mm^3) and their possible displacements of ± 5 mm. The distances in the scheme are given in millimeter. Right: Picture of the shim unit with the aluminum holders for each permanent magnet. Each aluminum holder allows for the individual displacement of a magnet.

of even further improving the homogeneity of the new Halbach magnet. It shall be mentioned that the new Halbach magnet was built by Dr. Peter Blümler.

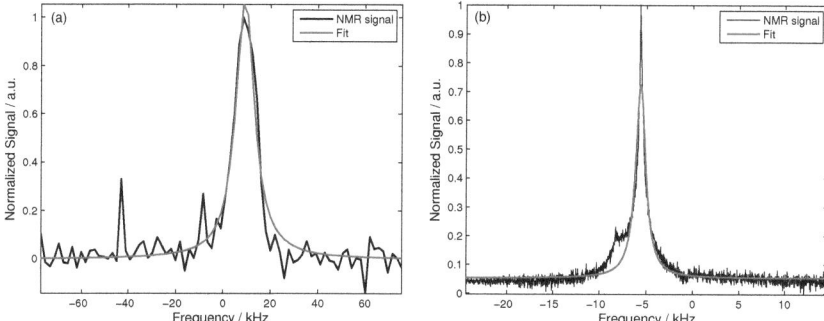

Figure 3.8.: Proton NMR signal in the old (left) and new Halbach magnet (right). (a) Fit of a Lorentzian line shape (red line) to the NMR signal in the old Halbach magnet yielding a full-width at half-maximum of approximately 10 kHz. (b) The NMR spectrum in the new Halbach magnet was fitted with a pseudo-Voigtian function (red line) to obtain a better fit. This fit yielded a full-width at half-maximum of less than 0.65 kHz.

3.4. DNP Probeheads

This Section covers the technical properties of the used and developed DNP probeheads which have to be at least double resonant. The double resonance is a necessary condition as for DNP experiments simultaneous microwave irradiation and NMR detection is required. Appropriately, a DNP probehead comprises an EPR cavity and an NMR coil. The development of double resonant probeheads requires a lot of expertise. Therefore, the technical realization of the home-built probeheads was done by the technicians Christian Bauer and Manfred Hehn. The obvious problem in the concept of a double resonance probehead is the technical realization of two different and decoupled resonators within one device. A possible design is the integration of the NMR coil within the EPR resonator which severely influences the quality factor of the EPR cavity. This approach was realized in the ENDOR (Figure 3.9) and CUBOID (Figure 3.10) probehead. To overcome the impact of the NMR coil, the probehead PH1004 (Figure 3.11) was built.

3. Technical Aspects of the Mobile Set-up

In this design the EPR and NMR mode are both optimized as the NMR coil is simultaneously the wall of the EPR cavity.

The EPR performance of the probeheads is characterized by the quality factor Q, the resonator volume V_c, the operating microwave frequency f_mw and the applied microwave power P_mw. The achieved B_{1e} amplitude within the resonator is related to Q, V_c, f_mw and P_mw via

$$B_{1e} = \sqrt{\frac{2\mu_0 Q}{V_c f_\text{mw}} \cdot P_\text{mw}} \ . \tag{3.2}$$

Hence, B_{1e} can be increased by decreasing the resonator volume and by increasing the microwave power or the quality factor. The microwave frequency is fixed as the probeheads are not tunable. After the construction of the probehead, only Q and P_mw can be varied by the experimentalist. As a result, the quotient $\sqrt{\frac{2\mu_0}{V_c f_\text{mw}}}$ is constant and often called the conversion factor c of the probehead which is characteristic for each EPR resonator. Consequently, Equation 3.2 takes the form

$$B_{1e} = c \cdot \sqrt{Q \cdot P_{mw}} \ . \tag{3.3}$$

To maximize the quality factor and implicitly B_{1e} for DNP experiments, the EPR probeheads were critically coupled.

The ENDOR Probehead

The ENDOR probehead (EN 4118X-MD-4) is originally designed for electron-nuclear double resonance (ENDOR) experiments. It was developed by Bruker and is commercially available. The design of the resonator is cylindric and the NMR coil is implemented within the resonator in the Helmholtz arrangement (Figure 3.9). The Helmholtz coil has to be aligned parallelly to the B_0 field in order to spatially decouple the B_{1e} and the B_0 field as it is indicated in Figure 3.9. The EPR active volume is approximately $4 \times 4 \times 4$ mm^3.[41] Outside this volume, the B_{1e} field drops to zero within a few millimeters. Accordingly, for an optimal DNP performance, only sample heights ≤ 4 mm must be used. Additionally, active cooling of the EPR resonator stabilizes the EPR mode during the microwave irradiation of a DNP experiment which increases the achievable NMR signal

3. Technical Aspects of the Mobile Set-up

enhancement.[41] The EPR properties of the ENDOR probehead are listed in Table 3.3 as well as for the home-built probeheads CUBOID and PH1004. The EPR mode of the ENDOR probehead is very sensitive to the water amount inside the cavity. Appropriately, only DNP experiments with capillaries and 3 mm tubes are feasible.

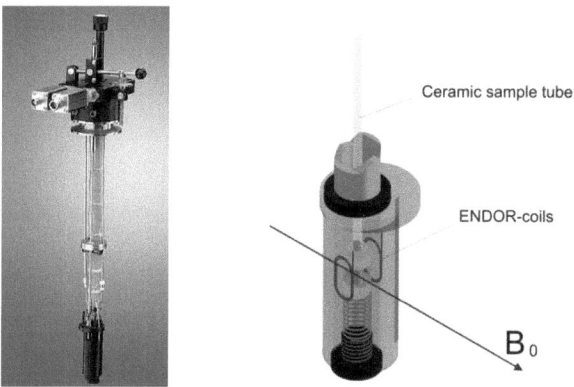

Figure 3.9.: Used EN 4118X-MD-4 probehead for ENDOR experiments with a cylindric design from Bruker. The static field B_0 must be perpendicular to the microwave B_{1e} field.

Table 3.3.: Properties of the tested probeheads. To determine the loaded quality factor, water samples were used.

Probehead	ENDOR (EN 4118X-MD-4)	CUBOID	PH1004
Design	cylindric	cuboid	cylindric
Resonator Volume [mm^3]	163	10,300	23,600
Unloaded Q	2282	476	460
Loaded Q (Capillary)	1801	476	536
Loaded Q (3 mm Tube)	334	368	-
f_{mw} [GHz]	9.7	9.76	9.45
Conversion Factor c [G/W$^{\frac{1}{2}}$]	~0.02	N.A.	N.A.
$f_{rf}(^1H)$ [MHz]	14.7	13.8 - 25	14.3

The CUBOID Probehead

The probehead CUBOID possesses a cuboid design and is shown in Figure 3.10. Consequently, it operates in TE_{102} mode with a resonance frequency of 9.76 GHz. The dimensions of the resonator are $23.2 \times 11.1 \times 40.0$ mm^3 so that in one direction a multiple of the microwave wavelength fits into the resonator. Here the solenoid NMR coil was integrated into the EPR resonator similar to the ENDOR probehead. The severe drawback of the coil's integration into the resonator is the deterioration of the EPR mode. The lowering of the quality factor (cf. Table 3.3) and its limiting effect are discussed in Chapter 4. The advantage of the solenoid NMR coil design is the automatic decoupling of B_0 and B_{1e}. The properties of the CUBOID are listed in Table 3.3, as well.

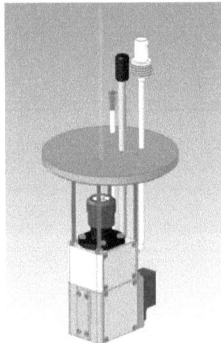

Figure 3.10.: Tested home-built rectangular probehead (CUBOID). Here, the NMR coil is cylindrical around the vertical axis. As a result, the external field is always perpendicular to the generated B_{1e} field.

The PH1004 Probehead

From the EPR point of view, the PH1004 probehead is a reconstruction of the ENDOR probehead with enlarged dimensions. The design of PH1004 and its dimensions can be taken from Figure 3.11. The difference to the ENDOR probehead refers to the construction of the NMR coil. It is not integrated in the EPR cavity. Actually, it is the EPR cavity which can be used to detect NMR signals. This design was proposed in our group by the diploma student Lasse Jagschies but first manufactured after his thesis.[58] The drawback of this design consists of the low filling factor of the NMR coil which

3. Technical Aspects of the Mobile Set-up

results in a very poor signal-to-noise ratio for small sample volumes, *e.g.* capillaries. Another feature of this probehead is the triple resonance so that not only proton spins but also carbon spins can be detected. The properties of PH1004 are summarized in Table 3.3, too. The DNP performance of the home-built probeheads as compared to the ENDOR probehead is discussed in Chapter 4.

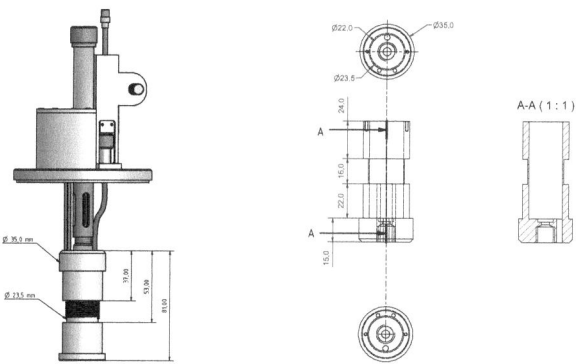

Figure 3.11.: CAD scheme (left) and technical drawing (right) of the probehead PH1004. The length of the cylindric resonator is 62 mm with a diameter of 22 mm. In the CAD scheme, the NMR coil is visualized and its length is approximately 16 mm (right).

3.5. Implementation of LabVIEW

Microwave irradiation of the DNP sample is essential for the performance of DNP experiments. To measure the NMR signal enhancement in dependence of the irradiation time, which yields a DNP build-up curve, the irradiation time has to be adjusted as accurately as possible. To that end, a LabVIEW program (National Instruments, LabVIEW 2010) was implemented by Christian Bauer to the set-up which allows for an automated control of the microwave source, the microwave amplifier and the NMR spectrometer. The graphical user interface of the program to manually operate the devices is shown in Figure 3.12.

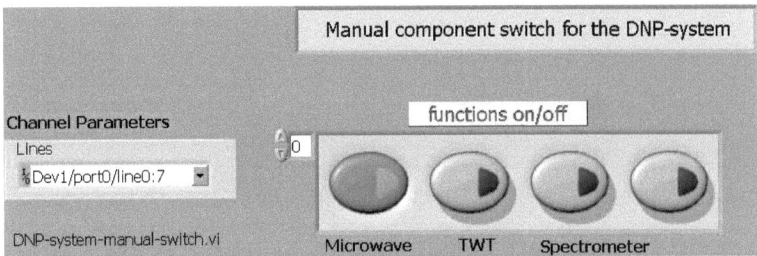

Figure 3.12.: Graphical user interface for the manual control of the microwave source, the microwave amplifier and the NMR spectrometer. Here, only the microwave source is switched on.

Supplementary to the LabVIEW program, the NMR pulse program had to be changed to allow for a defined triggering of the spectrometer. The program for adjusting the microwave irradiation time, which was used for the measurements in Section 7.3, is depicted in Figure 3.13. Without the automated experiments, the determination of the spin-lattice relaxation time, as presented in Section 7.3, is not possible.

3. Technical Aspects of the Mobile Set-up

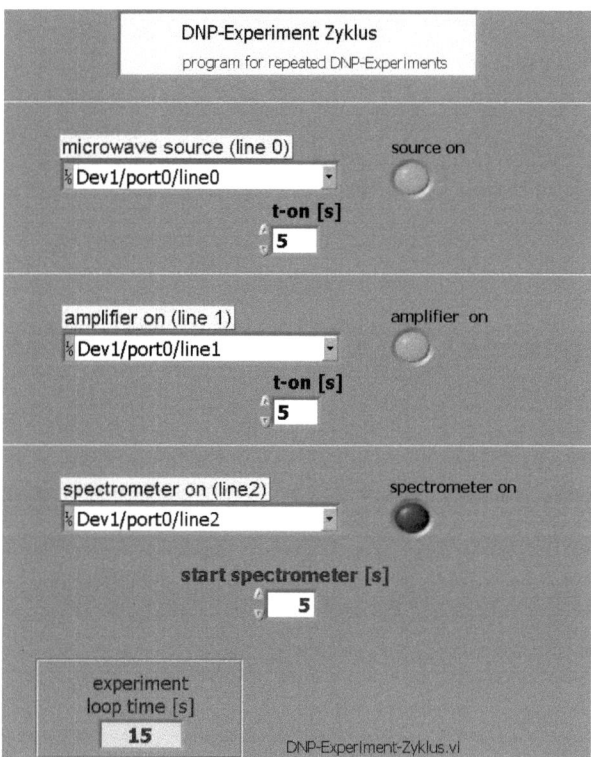

Figure 3.13.: Graphical user interface for the accurate adjustment of the microwave irradiation time. Here, the microwave irradiation time is adjusted to 5 seconds and the NMR acquisition starts exactly at the end of the irradiation. When the screen shot was taken, the microwave source and the TWT amplifier were switched on (bright green) whereas the NMR spectrometer was still waiting for the trigger signal (dark green).

4. DNP Performance of the Probeheads

All presented results in the following Chapters 5, 7 and 6 were obtained with the ENDOR probehead. Nevertheless, it reveals two weaknesses:

(i) Only small sample amounts can be polarized.

(ii) The probehead is only double resonant which hinders more sophisticated pulse sequences and experiments.

Property (i) makes the ENDOR probehead only conditionally suitable for medical applications as for this purpose larger sample amounts are required. Property (ii) restricts the investigation of radical-solvent systems to simple one pulse or cpmg-detection mode experiments which limits the insight into such a system. As a result, the home-built probeheads CUBOID and PH1004 were constructed so that larger sample amounts can be inserted. They were already introduced in Section 3.4. First, the probehead CUBOID was constructed to realize the polarization of larger sample amounts. Subsequently, the triple resonance should be implemented in a second step. This was realized with the probehead PH1004. The DNP performance of these home-built probeheads is presented and compared in the following Subsections.

4.1. CUBOID Probehead

For an easy comparison, the DNP performance of the CUBOID probehead was tested with the standard radical TEMPOL. The DNP enhancement of TEMPOL ($c = 4$ mM) in a capillary is plotted in Figure 4.1 (a) to display its absolute DNP performance. In Figure 4.1 (b), the well-known plot for the determination of the maximum achievable enhancement is projected. The measured DNP enhancement was up to $E = -51$ and its projected maximum enhancement $E_{\max} = -123$. One can derive from the difference in E and $E_{\max}$ that the sample could not be fully saturated. Evidently, as seen by Figure 4.1 (a), even at microwave powers as large as 20 W the enhancement does not reach a plateau. The ratio of $E/E_{\max}$ corresponds to a saturation of 41%. On the contrary, the observed DNP enhancement for the ENDOR probehead comes very close to the projected enhancement (TEMPOL, $c = 5$ mM) where $E = -127$ and $E_{\max} = -134$ are similar. This ratio is ~ 0.94 which corresponds to a total saturation of 94%. Accordingly, the saturation in the ENDOR probehead is more than twice as good as for the CUBOID probehead.

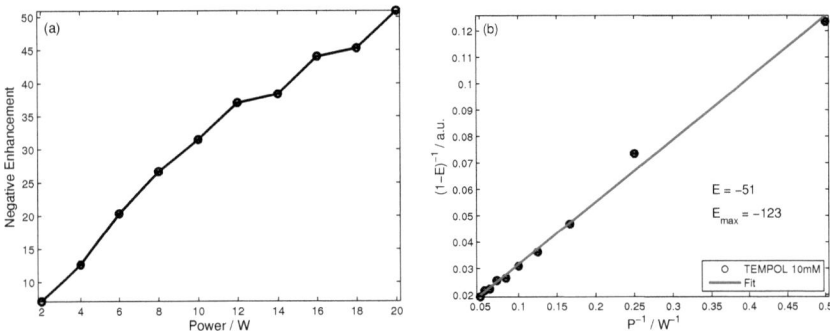

Figure 4.1.: (a) The plot shows the achieved enhancements at different microwave powers. (b) The same data points plotted as described in Section A.6 show the linear dependence which allows for an easy and fast calculation of the extrapolated enhancement $E_{\max}$.

Nevertheless, the CUBOID was designed for large sample amounts. Therefore, different sample tubes and shapes were used to test the NMR signal enhancement under these conditions. For convenience, these results are summarized in Table 4.1.

Table 4.1.: DNP performance of the CUBOID probehead for different sample amounts and tube shapes. The used sample was TEMPOL dissolved in de-ionized water and $c = 10$ mM.

Parameter \ Sample Tube	Capillary[a]	3 mm	Flattened 5 mm	5 mm	ENDOR 3mm
E	-51 ± 1	-39 ± 1	-37 ± 1	-14 ± 1	-43 ± 2
$E_{\max}$	-123 ± 1	-63 ± 2	-75 ± 3	N.A.	-48 ± 1

[a] TEMPOL with $c = 4$ mM

The measurements in the capillaries showed the best DNP performance and in the 5 mm tubes the worst. The measurements in the 3 mm tubes and the flattened 5 mm tubes gave similar enhancements. For all sample tubes, the saturation $E/E_{\max}$ ranged from 40 to 60% which is evident from Table 4.1 and the measured enhancements did not exceed $E = -51$. The largest sample tubes which can be inserted into the ENDOR probehead are 3 mm tubes. Consequently, the DNP enhancement of the ENDOR for 3 mm tubes was also measured and listed in Table 4.1. Remarkably, the extrapolated enhancement of the CUBOID is higher than the one for the ENDOR probehead.

Discussion

It is obvious that in the CUBOID probehead the samples can not be fully saturated even when applying high microwave power (*cf.* Figure 4.1). In the ENDOR probehead considerably lower microwave power ($P_{\text{mw}} \approx 4$ W) is sufficient to completely saturate the EPR lines. As a result, despite the larger extrapolated enhancement (3 mm tube), the actually achieved enhancement in the CUBOID lies below the one of the ENDOR probehead. It becomes even more apparent if one compares the results achieved in capillaries. The extrapolated enhancement ($E_{\max} = -123$, $c = 4$ mM) of the CUBOID is close to the one obtained for the ENDOR probehead ($E_{\max} = -134$, $c = 5$ mM, Table 5.2). On the contrary, the achieved enhancements differ by more than a factor of two ($E = -51$ and $E = -127$) which cannot be explained by the small concentration difference. The incomplete saturation indicates a poor conversion of the microwave power into an effective B_{1e} field. First of all, the integration of the NMR coil within the EPR cavity interferes with the EPR mode which severely lowers the quality factor Q. Indeed, Table 3.3 shows a much lower quality factor of $Q = 476$ for the CUBOID as compared to $Q = 1801$ for the ENDOR probehead when loaded with a capillary. The quality factors of the two probeheads approximate for a 3 mm sample tube (*cf.* Table 3.3). Along with equal Q values the achievable and extrapolated enhancements become similar (*cf.* Table 4.1). Moreover, the conversion factor $c \propto \frac{1}{\sqrt{V_c}}$ which is proportional to the inverse resonator volume is considerably lower for the CUBOID as their volumes differ by almost two orders of magnitude (*cf.* Table 4.1). These insights explain the incomplete saturation of $< 60\%$ even for high microwave power and are the origin of the low DNP enhancements.

4.2. PH1004 Probehead

To overcome the low quality factor, the PH1004 probehead was designed. Unfortunately, the quality factor could not be significantly increased by removing the NMR coil from the EPR cavity (cf. Table 3.3). Preliminary results of this probehead are summarized in Table 4.2. For this probehead, no errors are given as the signal-to-noise ratio was too low to specify reasonable error estimates. The low signal-to-noise ratio stems from the low filling factor of the NMR coil. The diameter of the NMR coil is approximately 22 mm (cf. Figure 3.11) whereas the inner diameters of a capillary and a 3 mm tube are < 1 mm and < 3 mm, respectively. The probehead PH1004 shows similar enhancement factors to the CUBOID but a lower extrapolated enhancement. Ergo, the achieved saturation is good but its potential DNP enhancement is worse than for the CUBOID. The triple resonance could not be tested, yet.

Table 4.2.: DNP performance of the PH1004 probehead for two different sample amounts. The used sample was TEMPOL dissolved in de-ionized water at $c = 10$ mM unless noted otherwise.

Parameter	Sample Tube Capillary	3 mm	ENDOR 3mm
E	-53	-37	-43 ± 2
$E_{\max}$	N.A.	-43	-48 ± 1

Discussion

The PH1004 probehead shows enhancements similar to the CUBOID and a good saturation. Nonetheless, it lacks the same inherent low conversion factor due to the large resonator volume and the low quality factor. Therefore, despite removing the NMR coil from the EPR cavity the maximum possible enhancement could not be improved yet. Beyond, the low signal-to-noise ratio, which is already observed for proton signals, might become a big obstacle if it comes to the detection of ^{13}C spins. Therefore, the realization of triple resonant experiments, e.g. carbon detection and simultaneous proton decoupling, might become extremely difficult.

4.3. Summary

The comparison of all three probeheads manifests the difficulties in constructing a good DNP probehead. The ENDOR probehead, which still shows by far the best DNP performance for capillaries, cannot be loaded with sample tubes larger than 3 mm. Additionally, a triple resonance for this probehead can not be constructed due to the low inductivity of the Helmholtz coil which is used for the NMR detection.

The first trial of a probehead which allows for the loading of large sample amounts was realized in the CUBOID. Yet, the DNP results display the difficulty of building a probehead which exhibits better properties than the ENDOR. For small sample amounts, the ENDOR is still better, for medium sample amounts, the probeheads are similar and, for large sample amounts, the enhancement for the CUBOID is so small that DNP experiments make no sense. The integrated NMR coil could be identified as the origin of the poor saturation (enhancement). To overcome this obstacle and to allow a triple resonance, the more sophisticated probehead PH1004 was designed. All needed properties are merged into this probehead:

(i) The NMR coil is implemented in a way to not disturb the EPR resonator.

(ii) The NMR coil was designed to allow for the implementation of a triple resonance.

(iii) Large sample amounts can be inserted into the EPR cavity.

All necesseties could be implemented but due to the large NMR coil the EPR resonator volume is large resulting again in a low conversion factor. Along with the low conversion factor, the NMR filling factor is very poor which makes it difficult to achieve a good signal-to-noise ratio for the NMR detection. Consequently, the DNP results could not be improved as compared to the CUBOID. The detection of carbon is also critical because of the low signal-to-noise ratio which is already observed for the proton detection.

To summarize, despite all efforts, the commercially available ENDOR probehead still shows the best DNP performance. As a consequence, all DNP experiments characterizing new polarizing agents were performed with the ENDOR probehead. Nevertheless, the CUBOID and the PH1004 feature the possibility of inserting large sample amounts. Beyond, the PH1004 probehead is capable of performing triple resonant experiments.

5. Overhauser-type DNP Performance of Polarizing Agents

DNP can be performed at high or low magnetic fields and at ambient or cryogenic temperatures.[12,40] Each working DNP mechanism has found its application.[13–15,41,56,59,60] Similar to the specialized applications under different experimental conditions, each DNP effect is maximized utilizing optimized radical systems. For example, triarylmethyl (TAM) radicals[61,62] and biradicals[39,60,63] are optimal polarizing agents for the solid-state DNP where the mechanisms *thermal mixing* and *cross effect* are working. Highly optimized radical systems are still missing for the Overhauser-type DNP at high as well as at low magnetic fields. Most experiments are carried out using stable nitroxide-based radicals[31,33,41,64–66] like TEMPOL which is still not completely understood from the DNP point of view. Consequently, a lot of work has to be done to optimize radical systems for applications utilizing the Overhauser-type DNP mechanism.

In this Chapter, two new radical systems are presented and their Overhauser-type DNP performance in an electromagnet is compared with the radical TEMPOL. The theory of the *Overhauser effect* is explained in Chapter 2.3. The polarizing agents are described in the Appendix B. The magnetic field strength for all experiments is 0.345 T. The Chapter is divided into several Sections in which the DNP efficiency of each polarizing agent is discussed starting with the well-known radical TEMPOL-^{14}N and ending with the most complex one which is a spin-labeled thermoresponsive polymer network (SL-hydrogel). The two synthesized polarizing agents are compared to the DNP performance of TEMPOL which can be seen as standard as it still yields the highest enhancement for the Overhauser-type DNP.

5.1. TEMPOL

The stable radical TEMPOL is introduced in Chapter B.1 in which the chemical structure is given, as well. It is a very common polarizing agent in Overhauser-type DNP experiments. In the following Section, the parameters of Equation 2.40 showing the achievable NMR signal enhancement are determined via DNP and DNP-independent measurements. The coupling factor ξ and the leakage factor f can be determined by nuclear spin-lattice relaxation measurements whereas the saturation parameter s can be estimated by EPR experiments and DNP measurements. Therefore, this Subsection is subdivided into an EPR part which characterizes the saturation parameter and a part which presents NMR and DNP results.

EPR Analysis

A typical CW EPR spectrum of ^{14}N-TEMPOL is shown in Figure 2.9. In Chapter 2.2 the origin of the three hyperfine lines is explained. From a CW EPR line shape analysis, the T_{2e} relaxation time can be estimated as it is introduced in Appendix A.2. The results of the CW EPR line shape analysis are summarized in Table 5.1. As the EPR lines broaden with increasing radical concentration, the T_{2e} times shorten with increasing concentration. This is visualized in Figure 5.1 in which the full-width at half-maximum (FWHM, $\Delta m_I(^{14}N) = 0$) is plotted for two different concentrations (a) and versus the radical concentration (b).

Table 5.1.: Parameters describing the EPR properties of TEMPOL at selected concentrations. The T_{2e} times are in perfect agreement with results from other publications.[41]

TEMPOL [mM]	0.5	2	5	10	20
HSE [MHz]	0.17 ± 0.02	0.66 ± 0.06	1.65 ± 0.15	3.3 ± 0.3	6.6 ± 0.6
T_{2e} [ns]	385	244	134	60	30

Supplementary, from Figure 5.1 (b), the HSE rate for TEMPOL can be estimated from the slope. The linear regression leads to a HSE rate per concentration unit of $k' = 0.33 \pm 0.03$ MHz/mM which can be seen in Figure 5.1 (b). This value serves as a basis for the determination of the HSE rates at all TEMPOL concentrations. The resulting HSE rates are summarized in Table 5.1.

The T_{1e} time of TEMPOL was determined via an inversion recovery experiment with FID detection. Unfortunately, the inversion recovery experiment could not be performed in an automated way via a pulse program due to a large background signal. To account for the background signal, for each delay time a FID on-resonant and a 100 G off-resonant was recorded. The background-free FID was obtained by subtracting the off-resonant from the on-resonant signal (cf. Appendix A.5). Subsequently, the area of the

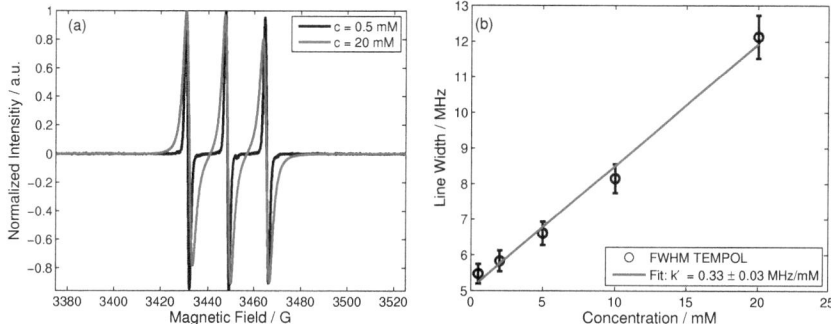

Figure 5.1.: (a) CW EPR spectra of TEMPOL for two different concentrations. The spectrum with the higher radical concentration obviously exhibits broader lines. (b) The FWHM line width of several concentrations is plotted versus the radical concentration which yields a straight line. The slope of the linear fit is used to estimate the Heisenberg spin exchange rate for a given concentration.

magnitude of the background-free FID was plotted versus the delay time which manifests an exponential increase (cf. Appendix A.5). The subsequent fit to this curve resulted in a T_{1e} time of 520 ns. The T_{1e} time was measured only for one concentration as T_{1e} does not depend on the radical concentration. Türke et al. measured for ^{15}N-TEMPOL a T_{1e} time of 350 ns which taking into account the differences between ^{14}N and ^{15}N is in agreement to the calculated value of 520 ns.

Overhauser-Type DNP Enhancement

The water proton NMR signal enhancement with dissolved TEMPOL radicals was measured many times. Here, only the highest enhancements achieved with the best equipment are presented for the investigated concentrations. The best available equipment for DNP consists of the electromagnet and the ENDOR probehead introduced in Chapter 3. Beyond the equipment, the enhancement is strongly influenced by the filling height of the capillary containing the sample as the B_{1e} field in the ENDOR probehead is only homogeneous and non-zero in a volume of roughly $4 \times 4 \times 4$ mm^3. Consequently, the presented results were achieved with a capillary filling height of only 3 mm compared to filling heights up to 10 mm which resulted in considerably lower enhancement factors. The highest observed enhancement was $E = -197 \pm 15$ ($c = 20$ mM, extrapolated $E_{\mathrm{max}} = -210 \pm 12$) for the measured concentrations at an adjusted temperature utilizing a closed cycle cryostat of $T = 15$ °C. Exemplarily, the microwave power dependence of TEMPOL ($c = 20$ mM) is visualized in Figure 5.2.

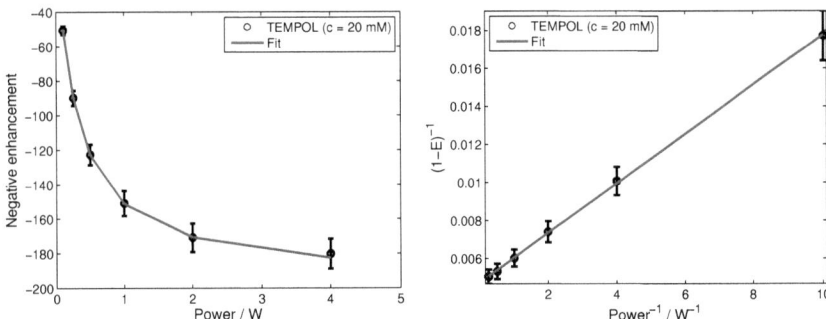

Figure 5.2.: The left figure shows the achieved enhancements at different microwave powers. The errors were calculated as described in Section A.6. The same data points plotted as described in Section A.6 show a linear dependence which allows for an easy and fast calculation of the extrapolated enhancement E_{max}.

All DNP related parameters of the different radical concentrations are summarized in Table 5.2. They were determined by T_{1n} and DNP measurements as explained in Section 2.3. All measured T_{1n} times at 0.35 and 7 T are listed in Table 5.3.

Table 5.2.: DNP parameters of TEMPOL at 15°C.

TEMPOL	E	E_{max}	f	ξ	s
0.5 mM	-33 ± 7	-35 ± 8	0.33 ± 0.06	0.35 ± 0.11	0.22 ± 0.06
2 mM	-66 ± 9	-68 ± 6	0.65 ± 0.03	0.34 ± 0.05	0.42 ± 0.07
5 mM	-127 ± 10	-134 ± 8	0.83 ± 0.02	0.32 ± 0.03	0.63 ± 0.10
10 mM	-153 ± 9	-156 ± 6	0.91 ± 0.01	0.42 ± 0.04	0.69 ± 0.11
20 mM	-197 ± 15	-210 ± 12	0.96 ± 0.01	0.44 ± 0.04	0.84 ± 0.13

Leakage Factor

From Equation 2.42, the leakage factor can be calculated for TEMPOL at different concentrations. By choosing a high radical concentration ($c \geq 10$ mM), the leakage

Table 5.3.: T_{1n} times of water protons for different TEMPOL concentrations.

TEMPOL	T_{1n} [s] (0.35 T)	T_{1n} [s] (7 T)
0 mM	2.3 ± 0.2	2.93 ± 0.1
0.5 mM	1.6 ± 0.1	2.23 ± 0.01
2 mM	0.80 ± 0.05	1.31 ± 0.01
5 mM	0.39 ± 0.02	0.65 ± 0.01
10 mM	0.21 ± 0.02	0.46 ± 0.01
20 mM	0.10 ± 0.01	0.25 ± 0.01

factor can be adjusted closely to unity. Table 5.2 shows that the leakage factor is $f \geq 0.9$ at radical concentrations of 10 mM. Consequently, the leakage factor is not the limiting factor in Overhauser-type DNP at appropriate radical concentrations as already explained in Chapter 2.3.

Coupling Factor and Saturation Parameter

The coupling factor derived from nuclear spin-lattice relaxation measurements gives values for the coupling strengths in the range from $\xi = 0.32$ to 0.44 (Table 5.2) which yields an average value of $\xi = 0.37 \pm 0.06$. Now, from the DNP and T_{1n} measurements and Equation 2.40, an effective saturation parameter s_{eff} can be retrieved which reaches values up to $s_{\text{eff}} = 0.84$. The value of the effective saturation parameter exceeds $s = 1/3$ which proves that with our experimental set-up we can effectively saturate more than one of the three hyperfine lines alone.

Discussion

The ^{1}H NMR signal enhancement of $E = -197$ for TEMPOL is one of the highest achieved enhancement for a non-deuterated TEMPO derivative published to this date. To understand the origin of this signal enhancement one has to scrutinize the parameters of Equation 2.40.

Leakage Factor

The leakage factor f is easy to access via T_{1n} relaxation measurements with very high accuracy. Its interpretation is simple as it gives the part of nuclear relaxation which is induced by the involved radicals. Additionally, f approaches its maximum value at relatively low concentrations ($c \geq 10$ mM). For this reason, f is not the critical component of the Overhauser-type DNP and subsequently plays a minor role.

Coupling Factor

The coupling factor ξ is determined independently from DNP measurements with nuclear relaxation measurements, as well. The coupling factor is not constant over the investigated concentration range as T_{1n} does not increase linearly with the magnetic field at different concentrations. This becomes clear from Table 5.2 in which ξ is increased at high concentrations. The discrepancy arises as one generally assumes a constant value of w_1 in Equation 2.51 and implicitly that $\mathcal{J}(\omega_I) \approx \mathcal{J}(0) = const.$ is valid. Therefore, the mean value and its standard deviation of ξ of Table 5.2 are taken as the correct value of the coupling factor for TEMPOL. The resulting coupling is $\xi = 0.37 \pm 0.06$ and reflects perfect agreement within error estimates with other publications[33,41] which report values from $\xi = 0.32$ up to 0.38. Up to the present, the coupling factor is the most difficult Overhauser parameter to determine as the theoretical derivation for the coupling factor by Hausser and Stehlik[40] becomes invalid at high magnetic fields. This can be seen in the deviation of the experimental data points from the theoretical curve in the high field limit (*cf.* Figure 12 of reference[40]). Relaxation theory predicts lower values for the coupling factor than observed. Armstrong *et al.*[64] proposed another model to obtain the coupling factor which yielded values of $\xi \sim 0.22$. This value is far too low and could therefore be discarded unambiguously by Türke *et al.*[41] Subsequently, the here reported coupling factors were always determined following the Hausser and Stehlik model.

Saturation Parameter

The saturation parameter s is computed most accurately performing pulse ELDOR experiments. Pulse ELDOR experiments require an EPR spectrometer which allows for sweeping the microwave frequency. The X-band spectrometer in our group is not capable of this experiment. As a consequence, the saturation factor is estimated theoretically using Equation 2.46 for non-coupled hyperfine lines assuming low radical concentrations in which HSE is negligible. On the other hand, from DNP measurements, an effective saturation parameter s_{eff} can be determined with the knowledge of the leakage and the coupling factor. The effective values of s range, besides the lowest concentration, above $s = 1/3$ verifying that more than one hyperfine line is saturated simultaneously. The highest saturation $s(c = 20 \text{ mM}) = 0.84$ is observed for the highest investigated concentration which emphasizes the effectiveness of HSE and the importance of appropriate radical concentrations for DNP. This total saturation of up to 84% verifies an almost full saturation of all hyperfine lines which is confirmed by Türke et al.[41] showing that the total saturation can be more than 80%. Interestingly, the saturation for $c = 0.5$ mM is only $s(c = 0.5 \text{ mM}) = 0.22$ which is less than the saturation of one hyperfine line. This observation indicates the suppressed effect of HSE at low radical concentrations and justifies to neglect spin exchange in Equation 2.46 in the low concentration region. The HSE rates depend linearly on the concentration. Thus, a reduction of the concentration from 20 mM to 0.5 mM implies a decrease of 40 of the effectiveness of HSE. Table 5.1 shows the linear decrease of the HSE rates from 6.6 MHz ($c = 20$ mM) to 0.17 MHz ($c = 0.5$ mM). Furthermore, the determination of the saturation factor and consequently the coupling factor as proposed by Armstrong et al.[64] can be discarded as this model assumes full saturation ($s = 1$) which leads to $\xi = 0.22$. Therefore, the experimental parameters of the Overhauser Equation are best determined in an independent fashion. It should be mentioned that the best enhancements were achieved in a set-up with an implemented cooling unit and a small sample volume. This cooling unit cannot avoid the severe heating of the sample but the cooling of the EPR resonator leads to a stabilization of the EPR mode which can result in a considerably larger enhancement.[41]

Conclusion

The work with ^{14}N-TEMPOL and the ENDOR probehead at the electromagnet has led to the observation of an enhancement of $E = -197$ of water protons at 15°C, the highest ever reported value for non-deuterated and ^{14}N radicals in solution. This enhancement was rationalized in terms of the Overhauser Equation and the results agree with the predictions from the available theory for EPR saturation.

The coupling factor as determined from T_{1n} relaxation measurements agrees within errors with nuclear magnetic resonance dispersion (NMRD) measurements from other groups[33,41] and only slightly exceeds the theoretical value as predicted by molecular dynamics simulations.[65] The enhancement of $E = -197$ verifies the exceptional performance of TEMPO derivatives for the Overhauser-type DNP. Moreover, it underlines the importance and effectiveness of HSE which can lead to an effective saturation far beyond the saturation of one hyperfine line.

Nonetheless, only high radical concentrations lead to these extraordinary large enhancements. Therefore, one problem of TEMPO derivatives remains: they are toxic for humans and animals in large concentrations if used for medical and biological applications and can influence the stability of biological molecules, *e.g.* proteins. Due to this problem, the need of biocompatible polarizing agents for medical applications and biological systems is obvious. Moreover, high radical concentrations imply NMR line broadening and shorten the life-time of the hyperpolarized state. Two possibilities to obtain *radical-free* and *non-toxic* hyperpolarized substances are the use of covalently bound radicals (*e.g.* to a matrix[67,68] or a responsive hydrogel[15]) and filtration via ion-exchange columns.[13]

5.2. Spin-Labeled Heparin

As mentioned in the introductory Chapter 1, the requirement of radicals for DNP experiments are the main drawback of this technique. For Overhauser-type DNP which is dominant at room temperature, nitroxide-based radicals are commonly used.[32,33] They are added to the sample as spin probes (cf. Section 2.3) yielding the highest observed enhancements, e.g. $E = -170$ at 0.35 T and $E = -29$ at 10 T at concentrations of about 20 mM.[41,66] However, for the investigation of biomolecules and proteins via DNP a lower radical concentration is beneficial because radicals lead to NMR line broadening and fast T_{1n} relaxation. Especially, when it comes to the medical application of hyperpolarized substances, the toxicity of stable free radicals plays an important role. Therefore, despite the undeniably good DNP performance of free nitroxide radicals, the need of biocompatible polarizing agents for medical applications remains an issue. In this context and following the purpose of this thesis, in collaboration with Dr. A. Kleschyov from the University of Mainz, the Overhauser-type DNP ($B_0 = 0.35$ T, room temperature) properties of spin-labeled heparins[69,70] (SL-heparin, Figure 5.3) with promising ^{1}H DNP enhancement factors and a high degree of biocompatibility are presented.

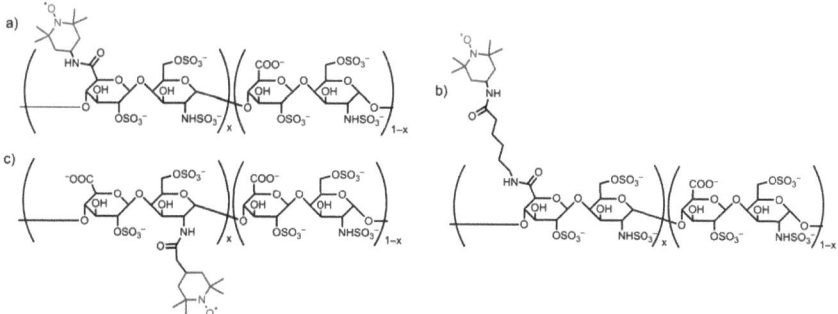

Figure 5.3.: Chemical structure of the disaccharide repeat unit of SL-heparin:
(a) Labeling via the carboxy group without linker (structure corresponds to SL-heparin 1 ($x = 0.18$) and SL-heparin 4 ($x = 0.72$)).
(b) Labeling via the carboxy group with a 6C-linker (structure corresponds to SL-heparin 2 ($x = 0.45$)).
(c) Labeling via the amino group without linker (structure corresponds to SL-heparin 3 ($x = 0.65$)).

The crucial point in the Overhauser-type DNP evaluation is the calculation of the saturation parameter and the coupling factor. The problem in determining ξ and s arises from Heisenberg spin exchange (HSE) or other residual couplings, *e.g.* dipolar interactions, when saturating one radical hyperfine line.[71] Here, the reported coupling factor was calculated using the well-established method by Hausser and Stehlik.[40] Subsequently, with the independently determined coupling strength and Equation 2.40 an effective saturation factor s_{eff} can be calculated. For three non-interacting EPR hyperfine lines (low radical concentration, no HSE and mixing via electron nuclear-spin relaxation) the saturation parameter depends on the electron spin T_{1e} and T_{2e} times and has the form of Equation 2.43:

$$s = \sum_{i=-1}^{1} \frac{1}{3} \cdot \frac{\omega_1^2 T_{1e} T_{2e}}{1 + \Omega_{0,i}^2 T_{2e}^2 + \omega_1^2 T_{1e} T_{2e}} \quad . \tag{5.1}$$

Equation 5.1 was derived from the Bloch Equations and ω_1 is the strength of the B_{1e} field. Here, $\Omega_{0,i} = \frac{B - B_{0,i}}{\gamma_e}$ with $i = 0, \pm 1$ denotes the off-resonance frequency of the individual EPR hyperfine line ($i = 1$ corresponds to the low field line, $i = 0$ to the center field line and $i = -1$ to the high field line) and γ_e the free electron magnetogyric ratio. Utilizing this Equation and Equation 2.40, the DNP enhancement can be calculated in dependence of the static magnetic field for uncoupled EPR hyperfine lines.

Results

CW EPR Spectra of SL-Heparins

In CW EPR spectra for free TEMPOL dissolved in water, we find relatively narrow EPR lines, *e.g.* the center line ($\Delta m_I(^{14}N) = 0$, FWHM= 0.20 mT, $c = 0.5$ mM), which can be simulated using a fast rotation model in the Redfield limit.[72] However, due to the binding of the radicals to the heparin backbone, steric restrictions apply and lead to incomplete spectral averaging. In addition, residual dipolar interactions (mainly at high labeling degrees) between the electron spins may emerge. Consequently, the center field line of the spectra of the SL-heparins show FWHM line widths ranging from 0.32 mT to 0.52 mT ($c = 0.5$ mM, Table 5.4). In the concentration and rotational motion regimes covered by our SL-heparins, dipolar and exchange interactions cannot be resolved or properly separated in the CW EPR spectra and lead to an underlying broadening of the lines (Figure 5.4).

5. Overhauser-type DNP Performance of Polarizing Agents

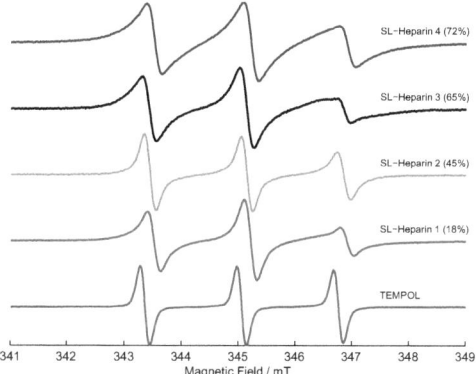

Figure 5.4.: CW EPR spectra of the SL-heparins and free TEMPOL at room temperature ($c = 0.5$ mM): SL-heparin 1 (18%, red), SL-heparin2 (45%, green), SL-heparin 3 (65%, black), SL-heparin 4 (72%, blue), TEMPOL (magenta). Note: The line width of SL-heparin 2 (45%, green) is narrower compared to the other SL-heparins due to the increased mobility of the spin-label (bond via 6C linker, *cf.* Figure 5.3 (b)).

Table 5.4.: Parameters of the SL-heparins ($c = 0.5$ mM) and TEMPOL ($c = 0.5$ mM) from CW EPR measurements. The denoted T_{2e} relaxation times were used for the calculation of the saturation profiles.

Radical	FWHM [mT] ($\Delta m_I = 0$)	effective correlation time [ns]	T_{2e} [ns]	k' [MHz/mM]
SL-heparin 1 (18%)	0.34 ± 0.02	~ 1	100 ± 10	0.076 ± 0.014
SL-heparin 2 (45%)	0.32 ± 0.02^a	~ 0.3	44 ± 8	0.050 ± 0.012
SL-heparin 3 (65%)	0.44 ± 0.02	~ 1.4	72 ± 8	0.046 ± 0.017
SL-heparin 4 (72%)	0.52 ± 0.02	~ 1	19 ± 3	0.036 ± 0.019
TEMPOL	0.20 ± 0.02	~ 0.03	385 ± 40	0.33 ± 0.3

[a] The line width of SL-heparin 2 is narrower than for SL-heparin 1 as the spin-label is bound to the heparin backbone via a linker which allows for higher mobility (cf. Figure 5.3 b).

The effective rotational correlation times $\tau_c \sim 1$ ns (Table 5.4) were determined by simulations using a model of uniaxial tumbling by Schneider and Freed as implemented in the MATLAB program package "Easyspin".[73] The CW EPR spectra were fitted with two rotational correlation rates accounting for the anisotropic, axial motion of the spin label (Table 5.4 for simplicity gives geometric mean approximation which is a good measure of the average correlation time $\tau_c = 1/\left(6 \cdot \sqrt[3]{D_{xx} \cdot D_{yy} \cdot D_{zz}}\right)$.[74] The electron spin-spin relaxation time T_{2e} was estimated by multiplying a non-broadened line shape reflecting the restricted motion with a stretched exponential decay function in the time domain (see Experimental Section). For the calculations of the T_{2e} times, negligible HSE rates were assumed as the HSE rates for all SL-heparins are much smaller than for TEMPOL of comparable concentration (Table 5.4).

The resulting T_{2e} times of the SL-heparins are in the range of tens of ns which is in good agreement with earlier publications for spin-labels.[75] These T_{2e} times are up to one order of magnitude shorter compared to those of free TEMPOL (Table 5.4). Furthermore, T_{2e} of SL-heparin shortens with increasing labeling degree which is evident from the line broadening visible in Figure 5.4. The T_{1e} time of TEMPOL was determined via an inversion recovery experiment with FID detection. The measured value of 520 ns agrees with values found in other publications.[76,77] The T_{1e} times of the SL-heparins could not be measured directly due to the very short transversal relaxation times but could be estimated for SL-heparin 4 from DNP measurements to be 190 ns (see the following Subsection). The Heisenberg spin exchange (HSE) rates per radical concentration were determined by fitting the slope of the CW EPR line width versus the radical concentration (Figure 5.5) yielding HSE rates for the spin-labels which are up to one order of magnitude lower than for TEMPOL (Table 5.4). The reason for the difference is the labeling to the heparin backbone. There is no strong overlap between individual, relatively stiff heparin chains at our concentrations. Thus, the spin label collision probability is drastically reduced by chemically attaching the radicals to the stiff chains.

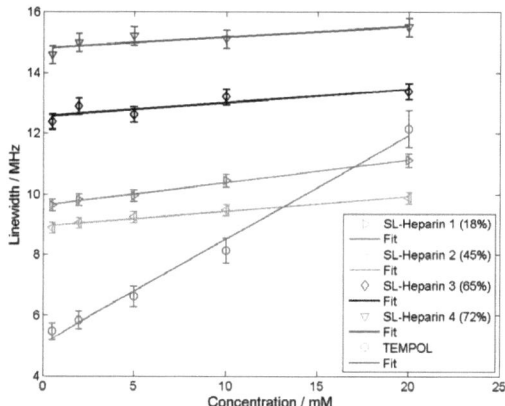

Figure 5.5.: Line width (FWHM) plotted versus the concentration. One sees the intrinsically broad lines of SL-heparin due to the labeling. Therefore, the slope is not as steep as for TEMPOL (magenta) and HSE is not as effective as for free radicals. The values for k' are given in Table 5.4.

DNP - ^{1}H Enhancements

The highest observed signal enhancements were $E = -110$ (SL-heparin 1, 18% labeling grade, $c = 20$ mM, extrapolated $E_{max} = -113$) and $E = -109$ (SL-heparin 4, 72% labeling grade, $c = 20$ mM, extrapolated $E_{max} = -113$) for the examined SL-heparins. For the reference radical TEMPOL ($c = 20$ mM) we could achieve an enhancement of $E = -112$ ($E_{max} = -119$). The maximum achievable enhancement E_{max} was computed by plotting the reciprocal Overhauser enhancement $(1 - E)^{-1}$ against the reciprocal power P^{-1} which should reveal a linear dependence (Figure 5.6, exemplarily plotted for SL-heparin 4). The intercept at $P^{-1} = 0$ of the linear regression yields the maximum enhancement when the EPR transition is saturated completely. The DNP parameters for all SL-heparins and TEMPOL are summarized in Table 5.5. Remarkably, for very low concentrations, the highly labeled SL-heparins 3 and 4 even showed a better DNP performance than TEMPOL. On the other hand, SL-heparin 1 and 2 with a low labeling degree and a long chemical linker, respectively, give lower enhancements than TEMPOL. In all DNP measurements, the power levels were up to $P_{\mathrm{max}} = 8$ W corresponding to a B_{1e} amplitude of 3.7 G for a loaded quality factor of the resonator of $Q = 1800$. The obtained B_{1e} amplitude is lower as compared to a publication by Türke et al.[41] as we did not cool the resonator actively which leads to a shift of the EPR mode.

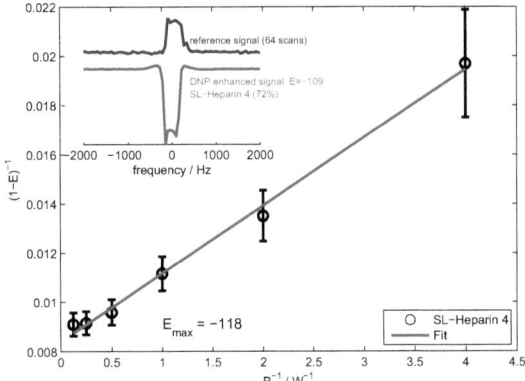

Figure 5.6.: Linear dependence of $(1-E)^{-1}$ from P^{-1} for SL-heparin 4 ($c = 20$ mM). Extrapolating P^{-1} yields the maximum enhancement E_{max}. The inset shows the best achieved enhancement for SL-heparin 4 at $c = 20$ mM.

Leakage Factor

From Equation 2.43, the leakage factor can be calculated for the SL-heparins at different concentrations. The part $(1-f)$ corresponds to the fraction of the total transition rates which are not induced by paramagnetic species. By choosing a high radical concentration ($c > 10$ mM), the leakage factor can be adjusted closely to unity. Table 5.5 shows that the leakage factor is $f > 0.9$ at radical concentrations of 20 mM and shows similar values for all measured radicals. Consequently, the leakage factor is not the limiting factor in

5. Overhauser-type DNP Performance of Polarizing Agents

Overhauser-type DNP at appropriate radical concentrations.

Coupling Factor and Saturation Parameter

The coupling factor is derived from nuclear spin-lattice relaxation measurements and gives values for the coupling strengths in the range from $\xi = 0.35$ to 0.44 (Table 5.5) which even surpass the coupling strength of free TEMPOL ($\xi = 0.39$). Now, from the DNP and T_{1n} measurements and Equation 2.40, an effective saturation parameter s_{eff} can be retrieved. At high concentrations, TEMPOL exhibits the highest saturation s_{eff} (TEMPOL, $c = 20$ mM)$= 0.48$ whereas SL-heparin ($c = 20$ mM) shows slightly lower effective saturations between 0.41 and 0.47. Contrary, in the low concentration limit ($c = 0.5$ mM), we find the best saturation for the SL-heparins 3 and 4 with a high labeling degree ($s_{\text{eff}} = 0.29$ and 0.26) which exceed the one of TEMPOL ($s_{\text{eff}} = 0.22$). The SL-heparins 1 and 2 with a low labeling degree and a long linker, respectively, showed the lowest saturations at this concentration. The values of the effective saturation parameter exceed $s = 1/3$ at high concentrations which proves that we can effectively saturate with our experimental set-up more than one of the three hyperfine lines alone.

Magnetic Field Dependence of the DNP Effect

In Figure 5.7 (a), the typical dependence of the DNP enhancement on the magnetic field strength is presented for TEMPOL. To this end, we plotted the first integral of the CW EPR spectrum and the DNP enhancement versus the magnetic field offset. For the measurement of the DNP profile, the EPR frequency was fixed and the magnetic field was swept. Consequently, the bandwidth of the resonator does not matter. We noticed that in between the EPR hyperfine lines the enhancement drops practically to zero and follows the EPR line although the DNP profile is already power-broadened. This means that the hyperfine lines are well-separated (which means that HSE is not apparent). The best enhancement under these experimental conditions ($c = 0.5$ mM, $P_{\text{mw}} = 0.5$ W) was $E = -15$. Figure 5.7 (b) shows the same plot for SL-heparin 4 featuring the highest labeling degree (72%) and the broadest EPR lines.

Contrary to the results depicted in Figure 5.7 (a), in between the hyperfine lines the enhancement does not drop below 60% of the maximum enhancement (Figure 5.7 (b)). Furthermore, one observes a discrepancy between the shape of the EPR line and the DNP profile which was not observed for TEMPOL. Here, under the same conditions the

Table 5.5.: DNP parameters of the SL-heparins and TEMPOL as obtained from DNP and T_{1n} measurements

Radical	c [mM]	E	E_{max}	f	ξ	s_{eff}
SL-heparin 1 (18%)	0.5	-15 ± 1	-17 ± 1	0.35 ± 0.06		0.17 ± 0.03
SL-heparin 1 (18%)	2	-44 ± 1	-52 ± 3	0.56 ± 0.04		0.29 ± 0.04
SL-heparin 1 (18%)	5	-76 ± 6	-85 ± 8	0.74 ± 0.02	0.43 ± 0.06	0.35 ± 0.05
SL-heparin 1 (18%)	10	-90 ± 5	-99 ± 6	0.89 ± 0.01		0.37 ± 0.04
SL-heparin 1 (18%)	20	-110 ± 8	-113 ± 6	0.94 ± 0.01		0.41 ± 0.05
SL-heparin 2 (45%)	0.5	-18 ± 1	-19 ± 2	0.28 ± 0.06		0.29 ± 0.07
SL-heparin 2 (45%)	2	-51 ± 6	-57 ± 8	0.57 ± 0.04		0.41 ± 0.05
SL-heparin 2 (45%)	5	-76 ± 8	-79 ± 6	0.74 ± 0.03	0.35 ± 0.04	0.44 ± 0.05
SL-heparin 2 (45%)	10	-92 ± 8	-97 ± 6	0.87 ± 0.02		0.47 ± 0.05
SL-heparin 2 (45%)	20	-103 ± 9	-106 ± 6	0.93 ± 0.01		0.47 ± 0.05
SL-heparin 3 (65%)	0.5	-24 ± 2	-27 ± 2	0.30 ± 0.06		0.29 ± 0.07
SL-heparin 3 (65%)	2	-61 ± 6	-69 ± 2	0.65 ± 0.03		0.33 ± 0.04
SL-heparin 3 (65%)	5	-85 ± 4	-95 ± 5	0.80 ± 0.02	0.44 ± 0.06	0.37 ± 0.04
SL-heparin 3 (65%)	10	-100 ± 3	-108 ± 3	0.89 ± 0.01		0.39 ± 0.05
SL-heparin 3 (65%)	20	-108 ± 4	-110 ± 5	0.93 ± 0.01		0.40 ± 0.05
SL-heparin 4 (72%)	0.5	-24 ± 2	-27 ± 2	0.37 ± 0.06		0.26 ± 0.05
SL-heparin 4 (72%)	2	-63 ± 5	-70 ± 3	0.63 ± 0.03		0.39 ± 0.06
SL-heparin 4 (72%)	5	-88 ± 6	-95 ± 4	0.81 ± 0.02	0.40 ± 0.05	0.41 ± 0.06
SL-heparin 4 (72%)	10	-97 ± 6	-105 ± 5	0.89 ± 0.01		0.42 ± 0.06
SL-heparin 4 (72%)	20	-109 ± 6	-118 ± 4	0.94 ± 0.01		0.45 ± 0.07
TEMPOL	0.5	-21 ± 2	-22 ± 2	0.38 ± 0.06		0.22 ± 0.04
TEMPOL	2	-60 ± 3	-64 ± 4	0.67 ± 0.03		0.36 ± 0.06
TEMPOL	5	-104 ± 8	-111 ± 6	0.83 ± 0.02	0.39 ± 0.06	0.49 ± 0.06
TEMPOL	10	-111 ± 4	-114 ± 3	0.91 ± 0.01		0.49 ± 0.06
TEMPOL	20	-112 ± 8	-119 ± 8	0.95 ± 0.01		0.48 ± 0.06

achieved enhancement is lower than for TEMPOL ($E = -13$). The same experiment was repeated with a significantly increased microwave power of 4 W. Therefore, the DNP profiles are obviously power-broadened and do not match a normal EPR line anymore. To account for this effect and the properties of the radicals, the DNP profile was calculated utilizing Equation 5.1 with the appropriate parameters from Table 5.4 and 5.5. For the calculation of the DNP profile, a correction factor of 0.5 had to be introduced to fit the measured DNP profiles of TEMPOL and SL-heparin 4, respectively. The correction factor was included as shown in the following Equation

$$E = 1 - \xi \cdot f \cdot s \cdot 658 \cdot \text{correction factor} \ . \tag{5.2}$$

5. Overhauser-type DNP Performance of Polarizing Agents

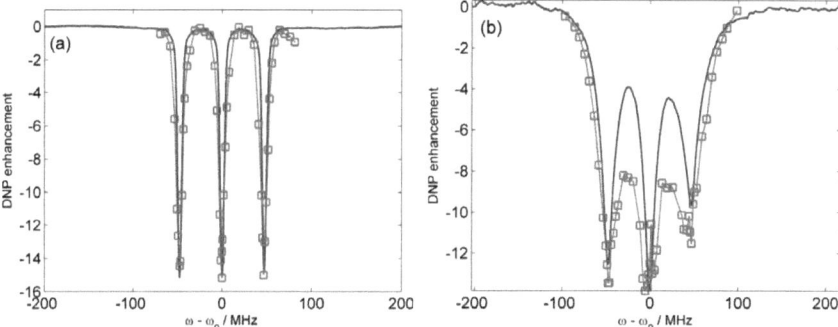

Figure 5.7.: Comparison of experimental CW EPR spectra (first integral) with ^{1}H DNP enhancement curves at room temperature. Solid lines depict the EPR spectra and squares the DNP enhancement as a function of the magnetic field offset (microwave power = 0.5 W). (a) The DNP enhancement between the hyperfine lines drops practically to zero for TEMPOL (c = 0.5 mM) and has the same shape as the EPR line. (b) Between the hyperfine lines the enhancement for SL-heparin 4 (72%, c = 0.5 mM) does not drop below 60% of the maximum enhancement.

The introduction of this correction factor which has no impact on our findings is justified by the fact that the filling height of the capillaries was $\sim$ 8 mm but the EPR cavity only had an effective volume of $4 \times 4 \times 4$ mm^3. For the calculation of the DNP profile of TEMPOL, the measured T_{1e} time (520 ns) was used. Subsequently, the magnetic field amplitude and the high field T_{2e} time ($T_{2e}(\Delta m_I=-1) = 300$ ns) were adjusted to fit the line shape. The low and center field line were implemented unchanged. The calculated DNP profile for TEMPOL together with the measured profile are presented in Figure 5.8 (a). Upon the power increase, the maximum enhancement increased slightly to $E = -17$ and in between the hyperfine lines the enhancement does not fall to zero anymore due to the power broadening. For the calculation of the DNP profile of SL-heparin 4, the magnetic field amplitude and the T_{1e} time were unknown as T_{1e} could not be measured (see above). The value for the magnetic field amplitude was determined from the DNP profile of TEMPOL as the experiments with SL-heparins were recorded under identical conditions. As a consequence, the T_{1e} time of SL-heparin 4 was adjusted to fit the DNP profile which is visualized in Figure 5.8 (b). As a result, the T_{1e} time of SL-heparin was found to be 190 ns. The T_{2e} times of the low ($T_{2e}(\Delta m_I=+1) = 15$ ns) and high ($T_{2e}(\Delta m_I=-1) = 10$ ns) field line also had to be adjusted as they obviously deviate from the T_{2e} time of the center field line ($T_{2e}(\Delta m_I=0) = 19$ ns). Figure 5.8 (b) visualizes that the low and high field line contribute a non-negligible part to the total saturation if one irradiates on-resonant at the center field line. The origin of this contribution is traced back to the broad lines caused by the short T_{2e} times. Accordingly, we found an enhancement of up to $E = -23$ which is significantly higher than the achieved enhancement of the DNP profile of TEMPOL.

Discussion

The attained enhancement $E = -112$ for TEMPOL has been reported recently in several publications[33,41] and does not include new information but demonstrates the reliability of our experimental set-up. The similar enhancements for SL-heparin of up to $E = -110$ validate the assumption that the electron-nuclear coupling is mainly of dipolar origin. The TEMPOL coupling factor as determined by T_{1n} relaxation measurements is reproduced within the error estimates as well.[33] It is slightly elevated as compared with recent publications.[33,78] The origin of the deviation can be traced back to the determination of w_1. w_1 is contrary to the assumption of Equation 2.51 field-dependent. As we determined w_1 at 300 MHz, the deviation stems from this measurement. The large errors for all coupling factors arise from the low signal-to-noise ration of capillaries at X-band field strengths. The coupling factors for the SL-heparins show similar values which slightly exceed the one for TEMPOL. Only the calculated factor for SL-heparin 2 (spin-labels bound via 6C linker) yields a lower coupling than for TEMPOL. The high coupling factors show that the SL-heparins are effective proton relaxation agents. One can safely assume that the polysaccharide heparin backbone is stretched rather than coiled. It was reported that heparin maintains its helical structure (at least in part) even in solutions[79] and the intrinsic charges lead to an additional stretching as is known from polyelectrolytes in general.[80,81] Consequently, the distance of closest approach between water molecules and spin-labels is almost unchanged as compared to free TEMPOL molecules. The diffusion of the water molecules around the heparin-attached spin labels should not be influenced too much, either. As the coupling factor is terminated primarily by the translational diffusion of the water molecules and the distance of closest approach, the use of Equation 2.52 for the calculation of ξ can be justified.

In Figure 5.4 it is evident that all SL-heparins have much broader CW EPR lines than the free radical TEMPOL. Obviously, with the bandwidth limitations well known in EPR, broad CW EPR lines are more difficult to saturate than narrow lines. Therefore,

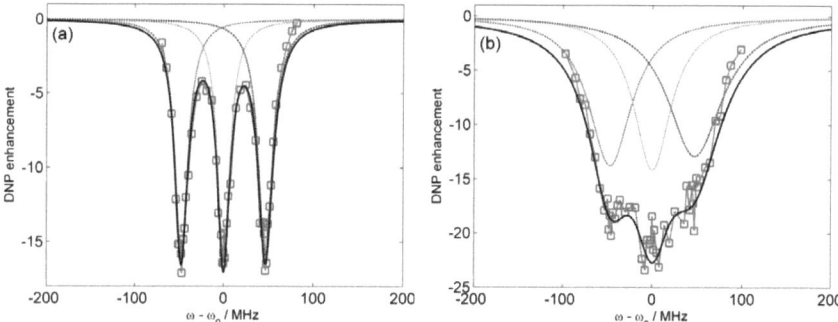

Figure 5.8.: Contribution of uncoupled EPR hyperfine lines to the DNP profile of TEMPOL (a) and SL-heparin 4 (b) as calculated from Equation 5.1 with the values from Table 5.4 and 5.5. Solid lines depict the calculated DNP profile and squares the DNP enhancement as a function of the magnetic field offset. The dotted lines show the contribution of each hyperfine line alone. (a) The uncoupled EPR hyperfine lines of TEMPOL show a negligible overlap and consequently the DNP enhancement between the hyperfine lines drops to -4. (b) Between the hyperfine lines, the enhancement for SL-heparin 4 (72%) does not drop significantly and follows the DNP profile as calculated from Equation 5.1. This effect shows the significant overlap of the uncoupled hyperfine lines for radicals with a short T_{2e} when one irradiates off-resonant. For the calculation of the DNP profile, different T_{2e} times for the hyperfine states were assumed as explained in the text.

a lot of research is conducted to develop radicals which display narrow EPR lines. Conventionally, one would expect enhancement factors (at the same incident power levels and at the same radical concentrations) which are considerably lower than those of free TEMPOL. However, it becomes apparent from our measurements that this is not true for the investigated SL-heparins. Actually, in the low concentration region where HSE is negligible the measured and extrapolated enhancements for the SL-heparins 3 and 4 (65% and 72% labeling degree, respectively) even surpass the DNP enhancement factor of free TEMPOL which can be seen in Table 5.5. For the explanation of these surprisingly high enhancements, we need to scrutinize the origin of the EPR line broadening in a qualitative way. The chemical binding of the nitroxides to the heparin backbone leads to a heterogeneous distribution and to an increased local concentration of electron spins as well as to a reduced mobility. The electron spin-electron spin relaxation time T_{2e} is decreased due to residual dipolar couplings and reduced mobility. To explain the effect of the change in the T_{2e} and T_{1e} times on the saturation we have to inspect the saturation factor as obtained from the Bloch Equations (cf. Equation 5.1). If we irradiate on-resonant, a simple Equation for one uncoupled EPR hyperfine lines is obtained:

$$s = \frac{1}{3} \cdot \frac{\omega_1^2 T_{1e} T_{2e}}{1 + \omega_1^2 T_{1e} T_{2e}} \qquad (5.3)$$

We find that a decrease in T_{1e} and T_{2e} yields a decreased saturation. Effectively, the on-resonant microwave irradiation of an uncoupled EPR hyperfine line alone leads to a significantly smaller saturation (enhancement) for the SL-heparins (cf. center field lines in Figure 5.8). On the other hand, it was mentioned by Sezer et al.[31] that even without being coupled, all hyperfine lines contribute to the total saturation.
This is seen from Equation 5.1 in which the off-resonant saturation is taken into account. The contribution from the off-resonant hyperfine lines can be calculated as described below. For of free TEMPOL (T_{1e} = 520 ns, $T_{2e}(\Delta m_I{=}{-}1)$ = 300 ns, $T_{2e}(\Delta m_I{=}0)$ = 385 ns, $T_{2e}(\Delta m_I{=}{+}1)$ = 385 ns, $B_{0,-1} \sim 1.69$ mT, $B_0 = 0$ mT, $B_{0,1} \sim -1.69$ mT, $B_{1e} \sim 0.27$ mT), the ratio $(\Omega_0^2 T_{2e}^2)/(\omega_1^2 T_{1e} T_{2e})$ is $\sim 23-29$ which means that the denominator is much larger than the nominator resulting in a contribution of only $\sim 7\%$ to the total saturation from the off-resonant hyperfine lines. This insight explains that the DNP enhancement drops significantly in between the hyperfine lines. With the simple Equation 5.1, the DNP profile of TEMPOL could be calculated as depicted in Figure 5.8 (a) with the parameters given above. For our SL-heparins (SL-heparin 4, T_{1e} = 190 ns, $T_{2e}(\Delta m_I{=}{-}1)$ = 10 ns, $T_{2e}(\Delta m_I{=}0)$ = 19 ns, $T_{2e}(\Delta m_I{=}{+}1)$ = 15 ns, $\Omega_0 \sim 1.69$ mT, $B_{1e} \sim 0.27$ mT), T_{2e} is an order of magnitude shorter and T_{1e} is also decreased so that $(\Omega_0^2 T_{2e}^2)/(\omega_1^2 T_{1e} T_{2e}) \sim 2-3$. This yields a larger contribution from the off-resonant EPR lines of up to $\sim 39\%$ to the total saturation. This contribution is again visualized by the significant overlap of the uncoupled hyperfine lines in Figure 5.8 (b). Therefore, we can saturate the SL-heparins 3 and 4 at $c = 0.5$ mM even better and achieve higher enhancements than for free TEMPOL (Table 5.5). This T_{2e} effect was already observed earlier by our group for another polyelectrolyte.[14] The mixing elevated rotational correlation times can be ruled out as the origin of the extra-saturation as SL-heparin 1 and 4 are identical systems which differ only in the labeling degree. Therefore, they must possess the same correlation time. If the extra-saturation effect stemmed from slowed-down rotational motion, SL-heparin 1 and 4 would have similar enhancements at $c = 0.5$ mM which is evidently not the case (cf. Table 5.5).
In the high concentration regime ($c = 20$ mM), all enhancements equalize as emerging

HSE rates compensate for the explained T_{2e} effect. Therefore, no trend at $c = 20$ mM for different lableing degrees is observed and all SL-heparins show similar enhancements (cf. Table 5.5). It is obvious that high HSE rates correlate with short T_{2e} times and vice versa (Table 5.4). The efficiency of HSE for SL-heparin is less pronounced than for TEMPOL as a result of the labeling. Furthermore, we note that the T_{1e} time of SL-heparin 4 as determined from the DNP profile is too short as compared to free radicals with similar rotational correlation times.[71] The origin of this shortening can be traced back to the fixed spatial distance of radical pairs generating a dipole-dipole interaction. Furthermore, Robinson et al. measured the T_{1e} time for a ^{15}N nitroxide radical. In addition, already slightly larger T_{1e} times would lead to much larger enhancements which is not in agreement with our measured DNP profile. Moreover, the average electron spin distance of SL-heparin 4 was measured independently via low-temperature EPR spectral fitting and double electron-electron resonance (DEER) spectroscopy. These experiments yield an average distance of < 1.5 nm (cf. Section 6.2) supporting our assumption of residual electronic dipole-dipole couplings.

As a result of our investigations, we propose that one can optimize the use of biological systems for DNP for which only small sample amounts are available by using heterogeneously distributed radicals on e.g. biological macromolecules. Such a system may be extremely effective for low concentrations in the biological agent due to the decreased T_{2e} time. The decreased T_{2e} time leads to a decreased on-resonant saturation but simultaneously to a non-negligible saturation from the off-resonant EPR lines. Thereby, the total saturation and consequently the enhancement can be increased in the limit of high microwave powers. This basic concept of a spin-labeled polysaccharide-system as polarizing agent should be applicable to most molecules. In this investigation, the spin-labeled macromolecules are biocompatible even at high concentration. This may be of great advantage for prospective in-vivo applications.

Conclusion

It was demonstrated that the SL-heparins show high ^{1}H DNP enhancements up to $E = -110$ ($E_{\max} = -118$). One striking result concerning the very broad lines of the SL-heparins in the EPR spectra is the fact that the achievable enhancements are comparable to free TEMPOL radicals. The enhancement of $E = -110$ proves directly that we are able to effectively saturate more than one EPR hyperfine line taking $f < 1$ and $\xi < 0.5$ into account. Usually, for free radicals broad EPR lines correspond to low achievable enhancement factors, due to incomplete saturation of the EPR lines. This effect may be due to an extremely effective total saturation with contributions from all hyperfine lines. The decreased electron spin-spin relaxation time could be identified as the reason for the high effective saturation. This was discussed qualitatively utilizing the saturation profile for uncoupled EPR hyperfine states as obtained from the Bloch Equations. An extension of our findings into a quantitative model must take into account electron spin-lattice relaxation (T_{1e}), electron nuclear-spin relaxation (T_{1ne}), electron spin-spin relaxation (T_{2e}) and Heisenberg spin exchange (HSE).

We propose that one can optimize the use of biological systems for DNP at physiological temperatures for which only small sample amounts are available by using heterogeneously distributed radicals by labeling e.g. biological macromolecules. Such a system may be extremely effective for low concentrations in the biological agent due to the decreased electron spin-spin relaxation time.

5.3. Thermoresponsive Spin-Labeled Hydrogel

DNP is a versatile polarization technique which utilizes several methods as introduced at the beginning of this Chapter. Each of these methods has its advantages and found applications, one severe problem, however, remains: stable and mostly toxic radicals have to be admixed to the target molecules. The radicals cause NMR line broadening if they are still present during the NMR detection and even more fundamentally they lead to fast T_{1n} relaxation immediately after the polarization step is completed. This severely limits the time frame during which the accomplished hyperpolarization can be used. This is a major problem for the *ex situ* DNP methods because of T_{1n} relaxation during the transfer time.

Hence, fast and reliable separation of radicals and polarized material remains an important issue for improving applicability of DNP. Two different procedures for radical separation were proposed in the literature: immobilization of the radicals in silica[67] or gel beads[68] combined with a continuous flow during the polarization step to separate the solute from the radicals, and the filtration of the radicals using ion-exchange columns.[13] Especially the application of gel beads manufactured from hydrophilic polymer networks seems to be promising: McCarney *et al.*[68] reported enhancement factors for water incorporated in a spin-labeled sepharose gel exceeding those obtained with silica-embedded radicals, which can be attributed to the higher mobility of the spin-labels in the water-swollen gels compared to the solid silica. In this work, for the first time, the use of thermoresponsive, spin-labeled hydrophilic polymer networks (SL-hydrogels)[82] for DNP is introduced. Supplementary, it is demonstrated that they allow fast and simple radical - solute separation (*cf.* Figure 5.9). The synthesis and the sample preparation are described in Chapter B.3. Heating of the swollen hydrogel to the critical temperature T_C induces a reversible, fast ($\leq$ 1s) and dramatic volume decrease ($\geq$ 500 volume%) thereby expelling water and other target molecules from the radical-bearing polymer network.[83]

In this work, two spin-labeled thermoresponsive hydrogels were synthesized (as described in detail in the supporting information) and investigated with regard to their applicability for Overhauser DNP. Two different labeling degrees were chosen for the study. In SL-hydrogel-1, up to 5% of all monomer units are labeled resulting in a radical concentration in the swollen state of $\sim$ 1 mM, while in SL-hydrogel-2 up to 15% of all monomer units bear a spin label representing an overall radical concentration of $\sim$ 6.8 mM below T_C, which is 63 °C in both systems[84] (*cf.* Figure 5.10). First, we characterize the system

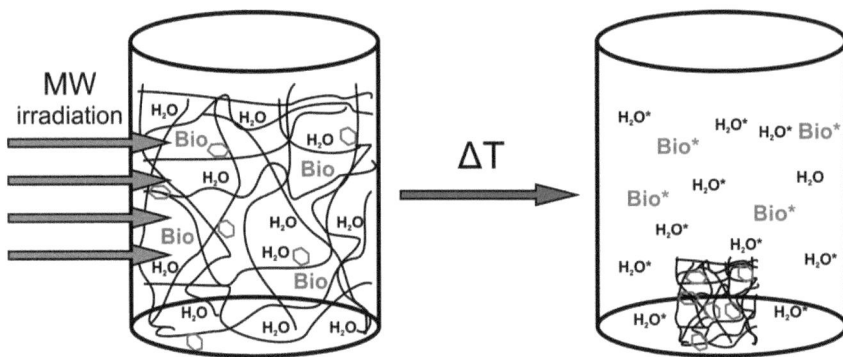

Figure 5.9.: Sketch of DNP polarization with a thermoresponsive spin-labeled hydrogel and subsequent temperature induced volume change resulting in the separation of the radicals and the hyperpolarized molecules (magenta hexagon: spin-label, Bio: target biomolecules, *: hyperpolarized).

by CW EPR. Then, the DNP achieved is demonstrated and the DNP parameters are determined. Finally, the reduction of DNP at elevated temperatures due to hydrogel collapse is shown.

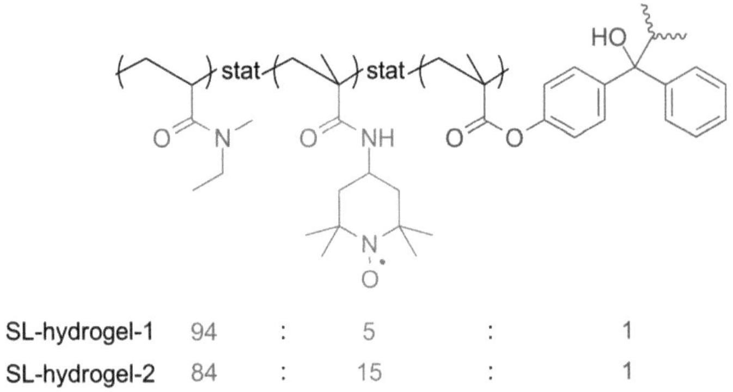

SL-hydrogel-1	94	:	5	:	1
SL-hydrogel-2	84	:	15	:	1

Figure 5.10.: Molecular structure of the two spin-labeled hydrogel networks. The part shown in green is responsible for the temperature-induced collapse, the red part allows for the spin-labeling and the blue part ensures the cross-linking of the hydrogel.

Overhauser-Type DNP Performance

CW EPR Characterization

CW EPR spectra were recorded in a temperature range from 5 to 55 °C and at $B_0 = 0.345$ T. In Figure 5.11, the CW EPR spectra of SL-hydrogel-2 are plotted for 5 and 55 °C, respectively. The spectrum of a typical nitroxide (TEMPOL, c=10mM) is plotted for comparison.

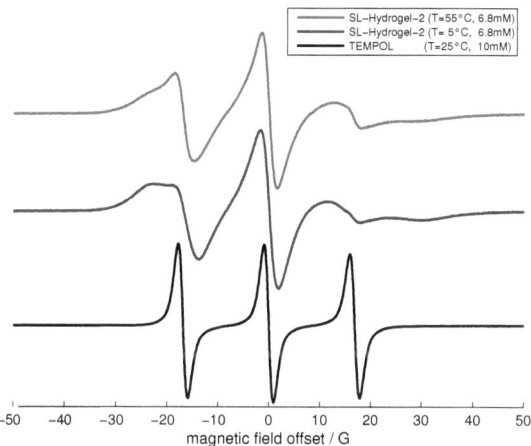

Figure 5.11.: X-band CW EPR spectra of SL-hydrogel-2 at 5 and 55 °C and the nitroxide-based free radical TEMPOL for comparison.

Obviously, the EPR lines of the SL-hydrogel are broader than those of free TEMPOL due to the restricted motion and mobility of the former where the radicals are covalently bound to the hydrogel network. Moreover, anisotropy in the rotational rates along and perpendicular to the labeling axis clearly manifests itself not only in the increased linewidths but also in additional features in the high- and low-field region of the spectrum. These additional peaks decrease when the temperature increases as the rotational correlation time reduces with rising temperature. The anisotropy and reduction in rotational motion could further be verified by spectral simulations (data not shown).

Characteristic DNP Factors and ^{1}H Relaxation Times

The highest observed ^{1}H DNP-enhanced NMR signal of water for our investigated SL-hydrogels is shown in Figure 5.12 (a). The used SL-hydrogels showed NMR signal enhancements of up to $E = -21.2 \pm 1.1$ (15 °C) and $E = -26.6 \pm 1.3$ (5 °C) at a mw power of 2 W for SL-hydrogel-1 and 2, respectively. To verify the reliability of our set-up, we measured the DNP enhancement of free TEMPOL ($c = 10$ mM) as $E = -148 \pm 8$ at an incident mw power of 2 W ($T = 15$ °C).

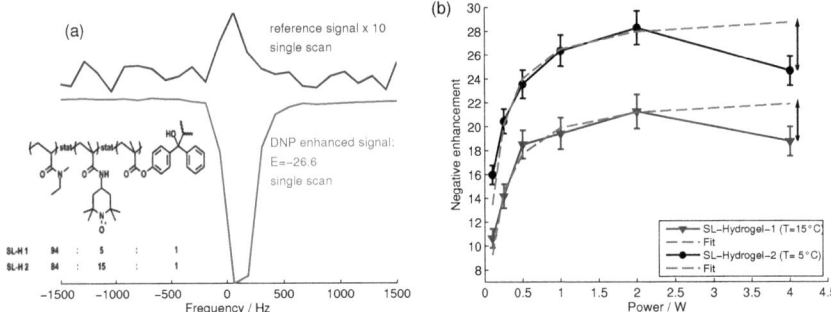

Figure 5.12.: Figure 5.12 (a) shows the best ^{1}H DNP-enhanced NMR signal for the SL-hydrogels (SL-hydrogel-2, $T = 5$ °C) and the chemical structure of SL-hydrogel 1 and 2. The corresponding monomer ratios are shown as an inset of Figure 5.12 (a). The measured enhancement is plotted against the power in Figure 5.12 (b) for both SL-hydrogels. The discrepancy between the expected and the actually measured signal enhancement at 4 W is indicated by the arrows. This effect is based on the collapse of the hydrogel network and is further explained in the text.

For microwave powers exceeding 2 W, we always observed a decline of the enhancement for all adjusted temperatures as can be seen from Figure 5.12 (b). This indicates that the applied mw irradiation on the sample can induce the fast thermal collapse (< 1 s) of the thermoresponsive SL-hydrogel due to heating of the water even in a cooled system. Exceeding T_C by microwave induced heating causes a collapse and an almost complete loss of water within the hydrogel network. In our case it is likely, though, that the samples are only partially collapsed due to local heating. Nonetheless, the spatial separation of the radicals and the solvent and solute (resembling a micro-phase separation) leads to a diminished coupling factor before the sample is completely saturated. This, in turn, explains the reduced enhancement for high mw power which are therefore slightly smaller than the enhancement McCarney et al. reported for ^{14}N spin-labeled sepharose gels ($E = -33$ at 6 W, $c = 10$ mM).[68]

Interestingly, the proton spin-lattice relaxation times in the hydrogel network, which are needed for the calculation of the leakage factor, were extremely short even without spin-labels ($\sim$ 480 ms) as compared to the relaxation time of free water ($\sim$ 2100 ms). For SL-hydrogels-1 and 2 we measured further reduced values of $T_{1n} \sim 330$ ms and $T_{1n} \sim 65$ ms, respectively. This behavior was observed before and is caused by the sterical confinement due to the hydrogel network and depends on the pore size of the network.[32] Small changes of the swelling degree lead to a marked change of T_{1n} as the degree of swelling severely affects the pore size and thus the spin-lattice relaxation time. Therefore, the uncertainties of the T_{1n} measurements were as large as 20%. From T_{1n} the leakage factor was determined as $f = 0.31 \pm 0.20$ (SL-hydrogel-1) and $f = 0.86 \pm 0.04$ (SL-hydrogel-2). The low leakage factor for SL-hydrogel-1 shows the impact of the reduced proton T_{1n} on the DNP efficiency. By the severe reduction of T_{1n} due to the hydrogel network the spin-labels are less effective polarizing agents. Only with an increased labeling degree f can be improved effectively. This is seen by the elevated leakage factor of SL-hydrogel-2. Furthermore, due to the large error estimated for T_{1n} and the difficulties to separate the saturation s and the coupling factor ξ for our system, we report the product of ξ and s. These are $\xi \cdot s = 0.108 \pm 0.068$ (SL-hydrogel-1) and $\xi \cdot s = 0.049 \pm 0.004$ (SL-hydrogel-2), respectively. The observed and calculated DNP parameters are summarized in Table 5.6 for the best achieved enhancements of the SL-hydrogels.

Table 5.6.: DNP parameters describing SL-hydrogel-1 at 15 °C and SL-hydrogel-2 at 5 °C.

hydrogel batch	Labeling degree [%]	estimated concentration [mM]	E_{max}	E	f	$\xi \cdot s$
SL-hydrogel-1	5	1.0 ± 0.2	-22.9 ± 1.2	-21.2 ± 1.1	0.31 ± 0.20	0.108 ± 0.068
SL-hydrogel-2	15	6.8 ± 0.5	-29.5 ± 1.5	-26.6 ± 1.3	0.86 ± 0.04	0.049 ± 0.004

As ξ is independent of the labeling degree, SL-hydrogel-1 is easier to saturate. The narrower linewidth of the EPR center line ($\Delta m_I = 0$) of SL-hydrogel-1 (FWHM = 13.4 ± 2.1 MHz) compared to SL-hydrogel-2 (FWHM = 17.2 ± 1.4 MHz) supports our conclusion. The higher enhancement in SL-hydrogel-2 solely results from its higher leakage factor. The difficulties in saturation also occur because (i) the radicals are immobilized in the network, so that Heisenberg spin exchange is suppressed; (ii) the mixing of the EPR lines by nuclear spin relaxation cannot be exploited as we are power limited due to the characteristic collapse of hydrogel network.

The thermal collapse of our presented system limits the achievable DNP, but it provides the possibility to combine the saturation of the EPR lines and the separation of hyperpolarized target molecules and toxic radicals in a single step. In addition, by the spatial separation after the thermal collapse, the spin-lattice relaxation time of the hyperpolarized molecules is prolonged up to the T_{1n} without radicals ($T_{1n} \sim 2100$ ms at 25 °C measured in water expelled from the hydrogel). This was verified by CW EPR, where no signal indicative of spin label was observed (data not shown).

Temperature-Dependent DNP Performance of SL-Hydrogels

Figure 5.13 shows the temperature dependence of the measured enhancement factors for both SL-hydrogels. The highest enhancements (for high mw power) were indeed achieved at low temperatures (5, 15 and 25 °C), while at elavated temperatures (45 and 55 °C) incomplete saturation and small dipolar coupling between electron and proton spins result from the partially collapsed hydrogel network. In a previous EPR study, Junk et al.[83] found that the thermally induced collapse on the molecular scale proceeds over a substantially broader temperature range than indicated by the sharp macroscopic volume transition at T_C.

A discontinuous collapse mechanism was suggested with a coexistence of collapsed and expanded hydrogel network region. This insight explains the smooth decrease in the DNP enhancement with increasing temperature in our data. Contrary to systems with no thermoresponsivity, in which the coupling factor and therefore the enhancement increases with increasing temperature, we found slightly reduced values for both parameters in the hydrogels. By reducing the mw power to 0.1 W - which is not enough to substantially heat the sample during microwave irradiation - we could separate the intrinsic DNP temperature dependence from the heating-induced collapse effect. Under these experimental conditions we found the expected rise of the DNP enhancement with temperature until the temperature approaches T_C (cf. Figure 5.13). To find the best enhancement factor, one has to compromise between two opposing effects. First, the "intrinsic" coupling factor increases with higher temperature due to the higher spin label rotational mobility (slightly narrowed EPR lines, see Figure 5.11). Second, the microwave heating-induced collapse of the hydrogel network efficiently separates the

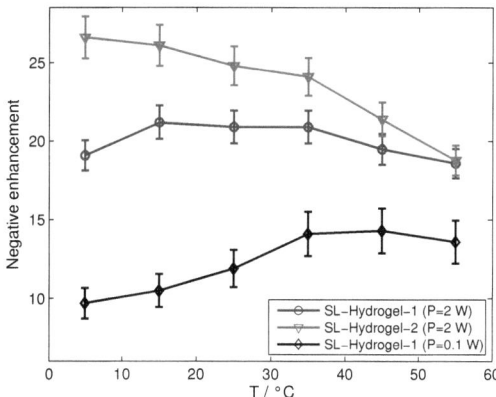

Figure 5.13.: Temperature dependence of the measured NMR signal enhancements for both SL-hydrogels (red and blue line) and the enhancement for SL-hydrogel-1 at 0.1 W (black line).

spin labels from water and potentially polarizable biomolecules resulting in lower DNP enhancements.

Conclusion

In summary, due to the collapse of the hydrogel at elevated temperatures, our presented system is characterized by a reduced T_{1n} during the microwave irradiation and an instantaneously prolonged T_{1n} afterwards and the benefit of a radical-free polarized sample. These unique properties make the studied system a good polarizing agent for *in situ* and *ex situ* DNP experiments. For *in situ* DNP methods the radical induced line broadening could be avoided and due to the prolonged T_{1n} time more complex 2D NMR experiments could be performed which might be beneficial for the investigation of biomolecules. Regarding the *ex situ* DNP methods the most obvious benefit lies in the prolonged lifetime of the hyperpolarization during the transport time thereby reducing the polarization loss until utilization of the hyperpolarized molecules. For application in a shuttle system which relies on the Overhauser DNP[33] the hydrogels have to be further optimized to yield maximum DNP enhancement at room temperature. However, the specified features make SL-hydrogel a very promising candidate for the dissolution DNP method. The obvious benefits are the prolonged T_{1n} after the dissolution step, the *radical-free* and *non-toxic* solute containing the hyperpolarized biomolecules allowing for biomedical applications. Furthermore, the severe heating effects which occur at room temperature are overcome at lower temperatures, *e.g.* $T \sim 1$ K. Then, the sample can be fully saturated. Preliminary low temperature DNP results at 0.345 T and 9 K show moderate proton enhancement factors which are discussed in Section 6.3. A magnetic field of 3.5 T and $T \sim 1$ K should result in a considerably larger NMR signal enhancement. Work along these lines in a collaboration with the Gruyter group at the EPFL in Lausanne is in progress.

5.4. Summary - Polarizing Agents

In the introduction to this Chapter the optimization of radical systems for the Overhauser-type DNP was proclaimed. Although TEMPOL still shows the best absolute DNP performance, the need of new redical systems was shown in the introductory parts of the Sections 5.2 and 5.3. The radicals cause line broadening and T_{1n} shortening to the NMR sample which limits the performance of Overhauser-type DNP and its applications. This can be overcome by removing the radicals from the sample.[15,68] Besides, the free TEMPO-based radicals are known to be toxic which hinders their use for medical applications. Therefore, the biocompatibility of new radicals or radical removal plays a key role.

In Chapter 5.2 the DNP results of the biocompatible polarizing agent SL-heparin are presented and discussed. The most remarkable insights of the experiments are the comparability of the DNP enhancements in the high concentration limit to free TEMPOL despite their very broad EPR lines. In the low concentration limit ($c = 0.5$ mM) for which HSE is negligible this line broadening can even lead to a better saturation of the EPR lines for high B_{1e} amplitudes. This T_{2e} effect which is caused by restricted rotational motion and dipolar couplings should be applicable to most molecules. Thereupon, we propose that one can optimize the use of biological systems for DNP at physiological temperatures for which only small sample amounts are available by using anisotropically distributed spin-labels on *e.g.* biological macromolecules.

The removal of radicals from a hyperpolarized sample can be done using ion-exchange columns[13] or by the immobilization of the radicals.[67,68] The presented thermoresponsive SL-hydrogels possess immobilized radicals which leads to a fast separation of the radicals and the hyperpolarized molecules upon microwave irradiation as it is visualized in Figure 5.9. The immobilization of the radicals leads to a broadening of the EPR lines. As a consequence, this radical system is more difficult to saturate as the line broadening is hardly caused by electronic dipole-dipole couplings as it was shown for SL-heparin. Moreover, the simultaneous separation of the radicals and the molecules upon a temperature increase due to microwave irradiation reduces the coupling factor. Including the filling height (5 − 8 mm, *cf.* Section B.3) and the low radical concentration, the moderate signal enhancements of up to $E = -26.6$ are reasonable. Nevertheless, due to the collapse of the hydrogel at elevated temperatures, our presented system is characterized by a reduced T_{1n} during the microwave irradiation and an instantaneously prolonged T_{1n} afterwards and the benefit of a *radical-free* polarized sample.

The next step to be done in the design of sophisticated polarizing agents, is the combination of the properties of SL-heparin and SL-hydrogel, which means *radical-free* hyperpolarized samples with a similar signal enhancement as obtained with free TEMPOL radicals.

6. Solid-State DNP Performance of Polarizing Agents

The success of a low-temperature DNP experiment largely depends on the EPR spectrum of the used radical. In low-temperature experiments, the glycerol-water solutions are not crystalline but a frozen glassy solution. For the purpose of magnetic resonance, a glass is a high-quality powder, more "homogeneous" in fact than a collection of small crystallites. In frozen crystallites, the dissolved radical will show a tendency to migrate to the surface, and under the influence of the magnetic forces the crystallites may be preferentially ordered. In an amorphous solid, the magnetic properties of the nuclei and radicals change. Theoretically, with decreasing temperature the T_{1n} time passes through a local minimum at the glass transition temperature ($T \sim 130$ K) and increases again when the temperature is further decreased. The T_{2n} time decreases monotonically with the temperature down to a few microseconds. Accordingly, the NMR lines are severely broadened due to emerging dipole-dipole interactions. The EPR line width broadens as a result of the temperature decrease whereas the T_{1e} time increases.

The important parameters of the solid-state DNP experiments are clarified in Equation 2.68. For convenience reasons, they are listed below:

(i) The electron dipole-dipole interaction frequency ω_D,

(ii) the electron spin-lattice relaxation time T_{1e},

(iii) the saturation parameter $W \cdot T_{1e}$,

(iv) the term $\frac{\Delta^2}{a\omega_D^2}$, which comprises the microwave offset frequency Δ and the exchange frequency of the electron spin-spin interaction reservoir ω_D and $a = W_{1D}/W_{1e}$,

(v) the nuclear spin-lattice relaxation time T_{1n} and

(vi) the interaction rate W_{nD} between the nuclei and the electron dipole reservoir.

To determine the parameters of the *thermal mixing* process in the list above, two different kinds of DNP measurements are performed:

(i) The DNP enhancement in dependence of the irradiated microwave power.

(ii) The DNP enhancement in dependence of the magnetic field which is called DNP profile.

For the parameter W, only the order of magnitude can be given due to the large errors of the measurements. The errors occur because of the line broadening and the accompanying decreased signal-to-noise ratio.

In this Chapter, the same polarizing agents as in Chapter 5 are scrutinized in respect to their DNP performance at cryogenic temperatures in an electromagnet. Consequently, the Chapter is divided into several Sections following the structure of the previous Chapter 5. Furthermore, the Sections are subdivided into a part which characterizes the EPR properties of the polarizing agents at low temperatures using ESE-detected spectra and DEER data and a DNP part which investigates the NMR signal enhancement. In all Sections, the DNP enhancement is compared to the DNP performance of TEMPO derivatives. All low temperature DNP measurements are performed at X-band and at $T \sim 9$ K which is the lowest achievable temperature of the closed cycle cryostat system. The solid-state DNP samples were filled in 3 mm tubes as frozen samples possess a low dielectric constant. Higher magnetic fields and lower temperatures are favorable for the achievable solid-state DNP polarization as the electron spin polarization reaches values close to unity ($B_0 = 3.5$ T and $T \sim 1$ K) but are not implementable in our set-up.

6.1. TEMPOL

The used TEMPO derivative TEMPOL is given in Appendix B where the chemical structure can be seen, as well. Trityl and nitroxide-based radicals are the standard polarizing agents for solid-state DNP experiments. In this thesis, exclusively nitroxide-based radicals were used for solid-state DNP experiments. For low-temperature DNP experiments much higher radical concentrations are needed to saturate the EPR line as compared to the Overhauser-type DNP mechanism. Therefore, the DNP performance of four samples containing high radical concentrations ($c = 10, 20, 40$ and 80 mM) are tested at X-band and $T = 9$ K.

EPR Results

For all concentrations, ESE-detected spectra were recorded at 50 K. In Figure 6.1, only the spectra of TEMPOL with the lowest and highest radical concentration are plotted. Obviously, the highly concentrated sample shows a broadened ESE-detected spectrum. The broadening is caused by electronic dipole-dipole couplings which do not cancel out in a frozen solid. The dipolar coupling strength is inversely proportional to the cubic distance of the electron spin pairs. Therefore, ESE-detected spectra of high concentrated samples are broader than samples with a lower radical concentration.

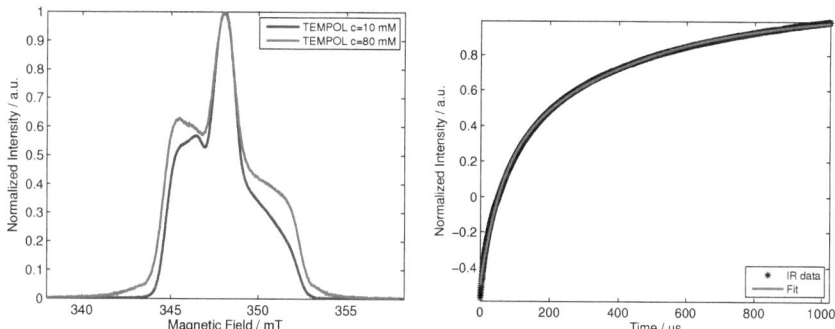

Figure 6.1.: Left: X-band ESE-detected spectra of TEMPOL at $T = 50$ K for $c = 0.5$ and 80 mM. Right: For example, the inversion recovery data and its bi-exponential fit are presented for $c = 20$ mM.

The electronic spin-lattice relaxation times were measured with an inversion recovery experiment at 9 K. The pulse sequence was $[\pi-\tau_1-\frac{\pi}{2}-\tau_2-\pi-\tau_2-\text{echo}]$ with subsequent integration over the whole echo. The resulting curves were fitted with a bi-exponential function as it is shown in Figure 6.1. Consequently, the obtained T_{1e} relaxation times comprise a fast and slow relaxing component. The acquired T_{1e} relaxation times are summarized in Table 6.1. The lowest TEMPOL concentration indicates the longest spin-lattice relaxation times.

Table 6.1.: Results of the bi-exponential fit to the inversion recovery data for all TEMPOL concentrations (cf. Figure 6.1).

TEMPOL [mM]	10	20	40	80
T_{1e} [μs] Fast Component	107.9 ± 2.1	54.5 ± 0.7	30.7 ± 0.3	17.0 ± 0.7
T_{1e} [μs] Slow Component	1099 ± 6	388 ± 5	309 ± 5	578 ± 42

^{1}H DNP Enhancement via Thermal Mixing

The electron dipole-dipole coupling frequency ω_D can be determined by measuring the DNP enhancement in dependence of the magnetic field. The distance of the two extrema in the DNP profile is proportional to ω_D. Exemplary for all concentrations, the electron dipole-dipole coupling frequency is calculated for a concentration of 40 mM. The corresponding measurements with their EPR line shape are given in Figure 6.2. The maximum of the EPR line is at $B_0 = 3479.3$ G for $\nu_{mw} = 9.77$ GHz. It follows that $\gamma_e/2\pi = 2.808$ MHz/G. The extrema of the enhancement are at $B_+ = 3498$ G and $B_- = 3448$ G which corresponds to a frequency of $\nu_+ = \frac{\gamma_e B_+}{2\pi} = 9.823$ GHz and $\nu_- = \frac{\gamma_e B_-}{2\pi} = 9.682$ GHz. The electronic dipole-dipole coupling frequency ω_D is determined according to Equation 2.69 to $\omega_D/2\pi = \frac{|\nu_+ - \nu_-|}{2} = 70.2$ MHz. The reasoning is summarized in the following Equation 6.1

$$\frac{\omega_D}{2\pi} = \frac{|\omega(E_+) - \omega(E_-)|}{2\pi \cdot 2} = \frac{|3498 - 3448| \text{ G} \cdot 2.808 \text{ MHz/G}}{2} = 70.2 \text{ MHz} . \quad (6.1)$$

This procedure was followed for each radical concentration. The results for ω_D are summarized in Table 6.2. In the DNP profile of Figure 6.2, the maximum enhancement

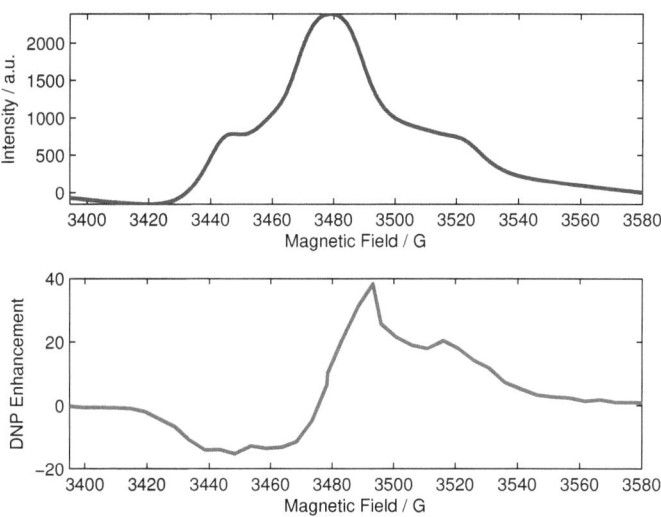

Figure 6.2.: X-band DNP profile of TEMPOL ($c = 40$ mM, bottom) and its corresponding ESE-detected EPR line shape (top) at 9 K.

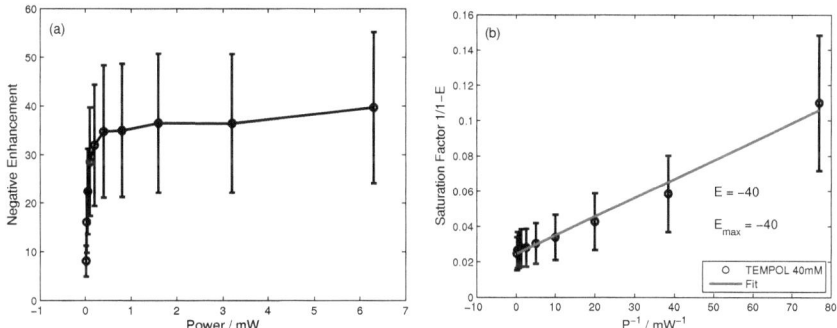

Figure 6.3.: (a) Power dependence of TEMPOL ($c = 40$ mM) at $B_0 = B_+ = 3498$ G and $T = 9$ K. (b) Plot of the inverse enhancement versus the inverse power to calculate the maximum enhancement as described in Appendix A.6.

was achieved at $B_0 = 3498$ G. The power dependence of the enhancement at $B_0 = B_+$ gives an estimation of W. Figure 6.3 shows this dependence for TEMPOL ($c = 40$ mM). In Figure 6.3 (a), one can clearly see that the sample is almost fully saturated for $P \geq 0.4$ mW which means that $E/E_{\max} \approx 1$. The determination of $E_{\max}$ as shown in Figure 6.3 (b) verifies this result. Consequently, the relation $\left(WT_{1e}\left(1 + \frac{\Delta^2}{a\omega_D^2}\right)\right) \gg 1$ is valid and subsequently according to Duijvestijn et al.[49] we can safely assume $\frac{\Delta^2}{a\omega_D^2} \approx 1$ and $W \gg 1$. A quantitative calculation of W for a microwave power which is nearly sufficient to saturate the sample ($P = 0.4$ mW) yields an exchange rate between the electron Zeeman and the dipolar electron reservoir of $W \approx 1.1 \cdot 10^8$ s^{-1}. Duijvestijn et al. determined a coupling rate of roughly $W \approx 1 \cdot 10^7$ s^{-1} which is one order of magnitude smaller than our results.

For the calculation of the interaction rate W_{nD}, the nuclear spin-lattice relaxation time is needed. T_{1n} was experimentally determined with a saturation recovery sequence in cpmg-detection mode. All measured T_{1n} times are shown in Table 6.2 which are in the range of a few hundred milliseconds. The measured and calculated parameters now allow for the computation of the interaction rate W_{nD} using Equation 2.68.

$$E = 1 + \frac{\gamma_e}{\gamma_n} \cdot W_{nD} \cdot T_{1n} \cdot \frac{\omega_n}{2\omega_D}$$
$$\Rightarrow 39.6 = 1 - \frac{-28.08 \text{ GHz/T}}{42.576 \text{ MHz/T}} \cdot W_{nD} \cdot 655 \text{ ms} \cdot \frac{14.873 \text{ MHz}}{2 \cdot 70.2 \text{ MHz}} \quad (6.2)$$
$$\Rightarrow W_{nD} = 0.84 \text{ s}^{-1} .$$

The obtained value of 0.84 s^{-1} has the same order of magnitude as the radicals used by Duijvestijn et al.[49] The computed W_{nD} values for the investigated concentrations are listed in Table 6.2, as well.

Table 6.2.: Summary of the important parameters of the *thermal mixing* effect for TEMPOL.

Concentration [mM]	ω_D [MHz]	E	T_{1n} [ms]	W_{nD} [s^{-1}]	W [s^{-1}]
10	42.1	30.1	1600	0.16	-
20	54.7	36.8	1100	0.36	$4.2 \cdot 10^8$
40	70.2	39.6	655	0.84	$1.1 \cdot 10^8$
80	70.2	49.0	69	10.0	-

The dependence of the *thermal mixing* effect on the concentration manifests itself in Table 6.2. The enhancement and the W_{nD} increase with increasing concentration whereas T_{1n} and ω_D decrease with increasing concentration. The spin-lattice relaxation time shows the highest sensitivity upon an increase of the radical concentration. As W_{nD} is calculated using the nuclear spin-lattice relaxation time, its variation shows the same order of magnitude as T_{1n}.

Discussion

The best enhancement of $E = 49.0$ (*cf.* Figure 6.4) implies a polarization transfer efficiency of $\sim 7.5\%$. The efficiency is low compared to recent results published by the group of R. Griffin in which they hyperpolarize their samples at high magnetic fields and high microwave power utilizing biradicals.[39,59,63] They achieve enhancements of $E \approx 200$ resulting in an efficiency of $\sim 30\%$. The difference in the enhancement originates on the one hand from the limited microwave power and on the other hand from the monomeric radicals we use. If one compares the DNP results under similar experimental conditions (similar microwave power and radicals) one can see that the obtained enhancements even exceed the ones published by the Griffin group.[59] Gerfen *et al.*[59] measured an enhancement of $E \approx 20$ which is more than a factor of two lower than the results presented here. The difference can be traced back to the differing magnetic fields as the achievable enhancement scales with $E \propto \frac{1}{B_0}$.

The concentration dependence of the *thermal mixing* is not as pronounced as for the *Overhauser effect* since the enhancement is proportional to $E \propto \frac{N_e^2}{\delta^2} \frac{B_{1e}^2}{B_0} T_{1n} T_{1e}$, where N_e is the number of electron spins and δ the EPR line width. Therefore, by increasing the radical concentration, the gain in enhancement due to N_e^2 is reduced by a broadened EPR line width δ and by decreased nuclear and electron spin-lattice relaxation times T_{1n} and T_{1e}. This effect can be seen in Table 6.2 where the enhancement increases only by 60% upon an 8-fold concentration increase. Accordingly, T_{1n} and T_{1e} decrease with the radical concentration as expected. The EPR line width broadens as expected upon the radical concentration which is presented in Figure 6.1.

Despite the fact that the microwave source is power limited, the connection of the electron Zeeman and the electron dipole reservoir was quite high as compared to other publications.[49] Figure 6.3 (a) represents the high exchange rate as the enhancement increases only marginally for microwave powers above 0.4 mW. The electron dipole-dipole

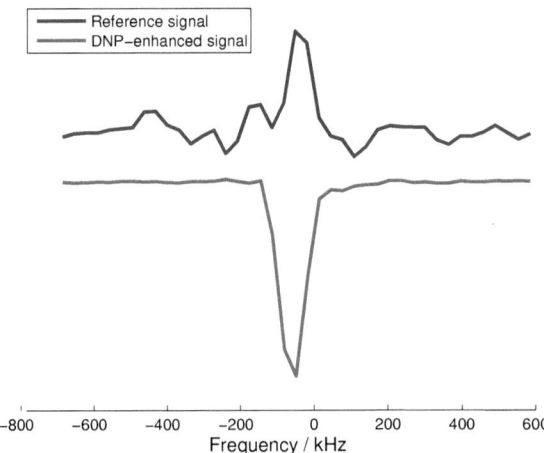

Figure 6.4.: Comparison of the reference signal with 100 accumulated scans to the DNP-enhanced signal with only four accumulated scans. The enhancement was approximately 40 and the microwave power 6 mW. The concentration is $c = 40$ mM. The DNP-enhanced signal was phase shifted by π for a better visualization.

interaction frequency ω_D as determined from the DNP profile increases with increasing concentration, as expected (*cf.* Table 6.2). The coupling rate W_{nD} between the dipolar electron and the nuclear Zeeman reservoir was calculated from DNP measurements using Equation 6.2. Due to the large deviations of the T_{1n} times, the W_{nD} rates showed the same deviations ranging from 0.16 up to 10.0 s^{-1}. Duijvestijn *et al.*[49] determined for a radical concentration of approximately 50 mM an exchange rate of 0.37 s^{-1} which shows the same order of magnitude as the measured 40 mM sample ($W_{nD} = 0.84$ s^{-1}). It is worth mentioning that the measured T_{1n} times (~ 1 s) are approximately two orders of magnitude shorter as compared to T_{1n} measurements in the high field of a nearly identical nitroxide in the same solvent (~ 100 s). This behavior was already observed earlier by our group.[58] It could not be determined yet if this shortening originates only from the magnetic field dependence of the nuclear spin-lattice relaxation time.

Finally, the observed enhancement of $E = 49$ reduces the acquisition time to accumulate the same signal-to-noise ratio by a factor of ~ 2400. This is visualized in Figure 6.4 where 100 accumulated reference scans (no microwave irradiation) are plotted together with a DNP-enhanced signal with four accumulated scans for a concentration of 40 mM.

6.2. Spin-Labeled Heparin

Many DNP research groups were established in the last decade due to the need and prospect of NMR signal enhancements, especially for medical applications. However, most used and investigated free radicals or spin-labeled molecules show high enhancements but are known to be harmful for animals and humans at the same time and have therefore to be removed from the hyperpolarized solution before utilization.[13,67,68] In this Section, the solid-state EPR and DNP properties of spin-labeled heparins[69] (SL-heparin, Figure 5.3) at cryogenic temperatures with promising ^{1}H DNP enhancement factors and a high degree of biocompatibility are reported. This spin-labeled macromolecule, which is a well-known anticoagulant, is biocompatible and can bind to the endothelium and various proteins.[85] The synthesis and MRI properties of these heparins (denoted as SL-heparins) are presented in the paper by Kleschyov $et~al.$[69,70] EPR methods can be used to measure distances from electron spin pairs by relating the measured dipolar coupling strength to their spatial separation. The crystal structure of heparin as determined by x-ray scattering is found in Figure 6.5[86] and reveals the helical structure of heparin. The results obtained by EPR experiments are used to analyze and interpret the DNP experiments. This Section is divided into an EPR part which investigates and compares the distance distributions as obtained by EPR pulse methods with the crystal structure of heparin as obtained by x-ray experiments. The second part analyzes the DNP properties of SL-heparin at cryogenic temperatures.

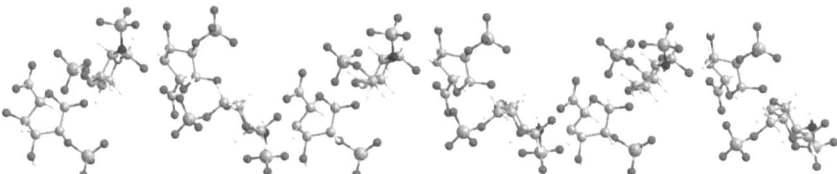

Figure 6.5.: Conformation of heparin: heparin helix having 12 saccharide units with sulfur atoms in yellow, nitrogen atoms in blue, oxygen atoms in red, carbon atoms in grey and hydrogen atoms in white (protein database (PDB) accession number for the coordinates are 1HPN).

Pulse EPR Results

ESE-Detected EPR Spectral Analysis

In ESE-detected spectra, the EPR line broadening effect is mainly caused by dipolar interactions as the motion of the molecules is frozen on the EPR timescale. For a quantitative analysis of the spectra, we fit with a simulated EPR spectrum that can account for the EPR powder spectrum and the dipole-dipole couplings of the electron spins by convolving a line shape that only reflects the EPR powder spectrum with a dipolar broadening function. We minimized the sum squared residual between the convolved and experimental spectrum as described in Appendix A.3. The evaluation of the data is demonstrated in Figure 6.6 for all SL-heparins and a typical TEMPOL spectrum is shown for comparison, as well.

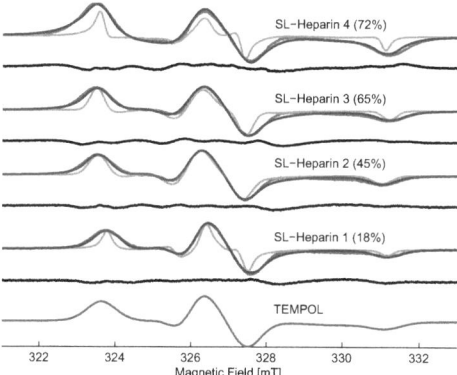

Figure 6.6.: ESE spectra of the SL-heparins together with their best fits according to a convolution approach explained in the text (convolved spectrum = uncoupled line shape ⊗ dipolar spectrum). Magenta: Experimental spectra, green: uncoupled line shape, blue: convolved spectra, black: residuals.

The resulting dipolar spectra, which consist of a superposition of Pake patterns, allow for the determination of the mean distance and the dipole-dipole coupling frequency ν_{dd} between the electron spins. For example, the mean distance of the spin labels for the SL-heparin 3 was found to be $d = 1.08$ nm ($\nu_{dd} = 41.3$ MHz). One has to keep in mind that the extracted distances overestimate short distances as the ESE spectra analysis is only sensitive to roughly 2 nm. All calculated distances and coupling frequencies are given in Table 6.3. Obviously, the higher the labeling degrees are, the higher the dipolar coupling frequencies become.

Table 6.3.: The second and third column show the radical-radical distances and the corresponding dipolar coupling frequencies from ESE-detected EPR measurements, respectively (cf. Figure 6.6).

Radical	mean radical pair distance [nm]	ν_{dd} [MHz]
SL-heparin 1 (18%)	1.76 ± 0.29	9.6 ± 1.6
SL-heparin 2 (45%)	1.06 ± 1.06	43.7 ± 43.7
SL-heparin 3 (65%)	1.08 ± 0.88	41.3 ± 33.7
SL-heparin 4 (72%)	0.91 ± 0.91	69.1 ± 69.1
TEMPOL[a]	5.6	0.3

[a] The radical-radical distance from monomeric TEMPOL was derived from its concentration ($c = 10$ mM).

DEER Analysis

Due to the statistical nature of the labeling, we expect Gaussian distance distributions at distinct distances when measuring the distribution of spin labels on the nanoscale. In DEER data, short electron-electron distances manifest themselves in fast modulations in particular in the beginning of the time domain signal and yield large dipolar coupling frequencies. From Figure 6.7, the fast signal decay within the first microsecond for the highly-labeled heparin (SL-heparin 4 (72%)) compared to the heparin with low labeling degree (SL-heparin 1 (18%)) can be seen.

The dipolar spectra after subtraction of the background are shown in Figure 6.8. The dipolar spectra can be simulated by a sum of pake patterns which are overlaid on the measured spectra. From these simulations, we can derive the occurring electron-electron spin coupling frequencies and their corresponding distances by using Equation 2.33. The extracted frequencies and distances are listed in Table 6.4.

The combined distance distributions extracted from DEER and ESE analysis are plotted in Figure 6.9. Distance approximations from ESE-detected EPR measurements are reliable to distances up to 2 nm whereas DEER measurements yield meaningful results down to 1.5 nm. Therefore, we plot the distances above 1.5 nm from ESE in dashed

6. Solid-State DNP Performance of Polarizing Agents

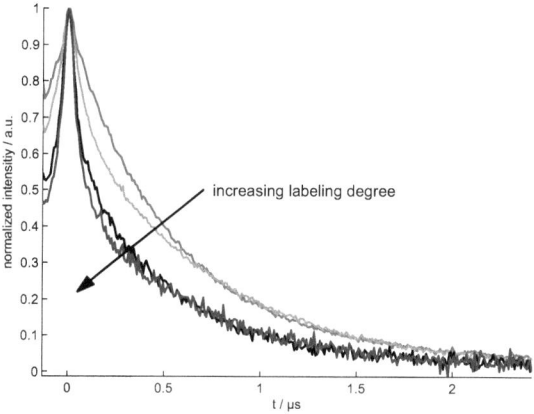

Figure 6.7.: Original DEER time traces of the four spin-labeled heparin macromolecules. SL-heparin 1 (18%, red), SL-heparin 2 (45%, green), SL-heparin 3 (65%, black), SL-heparin 4 (72%, blue). The data clearly reflect the increasing labeling degree in the faster decay of the signal at very short dipolar evolution times.

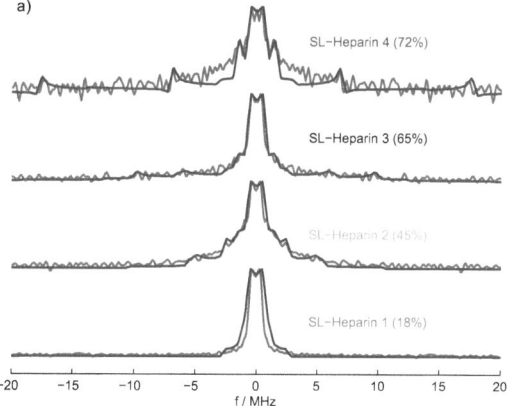

Figure 6.8.: Dipolar spectra of the DEER measurements: Magenta: Experimental spectra, blue: Pake patterns. The determined frequencies and their corresponding distances are listed in Table 6.4.

Table 6.4.: Summary of the determined frequencies and distances from the dipolar spectra in Figure 6.8.

Radical	$\nu_{dd,1}$ [MHz]	distance 1 [nm]	$\nu_{dd,2}$ [MHz]	distance 2 [nm]	$\nu_{dd,3}$ [MHz]	distance 3 [nm]	$\nu_{dd,4}$ [MHz]	distance 4 [nm]
SL-heparin 1 (18%)	-	-	-	-	1.29	3.43	0.41	5.03
SL-heparin 2 (45%)	5.24	2.15	2.76	2.66	1.36	3.37	0.41	5.03
SL-heparin 3 (65%)	9.71	1.75	6.22	2.03	1.38	3.35	0.41	5.02
SL-heparin 4 (72%)	17.8	1.43	7.02	1.95	1.52	3.25	0.41	5.02

lines to guide the eye as the data in this distance regime become less reliable than those from DEER results. Figures 6.8 and 6.9 clearly indicate that the conformation of the spin-labeled molecules and their labeling degree determine the spatial distribution of the spin-labels and hence also the dominant dipolar couplings between them.

The intensities in Figure 6.9 are not to scale but we find that the most prominent distance occurring for the spin-labeled heparin with the lowest labeling grade (SL-heparin 1 (18%)) lies at about 5 nm. For the other molecules, the prominent distances appear at much lower values of about 1 to 3 nm reflecting the higher labeling degree. Interestingly, all SL-heparins feature a peak at approximately 5 nm which might be an artifact that may arise from the background subtraction if the measured tracks are not yet decayed to a plateau value at the end of the measurement. Table 6.5 shows the distance values of the three distinct peaks in Figure 6.9 for the DEER data.

Table 6.5.: Summary of the fitting results of the three distinct peaks in the distance distribution (Figure 6.9). The peaks were fitted by a Gaussian line shape.

Radical	peak 1 [nm]	$\nu_{dd,1}$ [MHz]	peak 2 [nm]	$\nu_{dd,2}$ [MHz]	peak 3 [nm]	$\nu_{dd,3}$ [MHz]
SL-heparin 1 (18%)	-	-	3.67	1.05	5.15	0.38
SL-heparin 2 (45%)	2.36	3.96	3.68	1.04	5.18	0.37
SL-heparin 3 (65%)	1.87	7.96	3.11	1.73	4.98	0.42
SL-heparin 4 (72%)	<1.5	>15.42	3.85	0.91	4.82	0.46

Discussion

Heparin molecules are macromolecules with a disaccharide repeat unit.[86] The secondary structure of heparin is a helix and is maintained at least in part even in solution.[79] The translation per disaccharide along the helix axis was estimated by Yamaguchi et al.[79] to be 0.75-0.87 nm.[87] The crystallographic repeat unit consists of a tetrasaccharide and

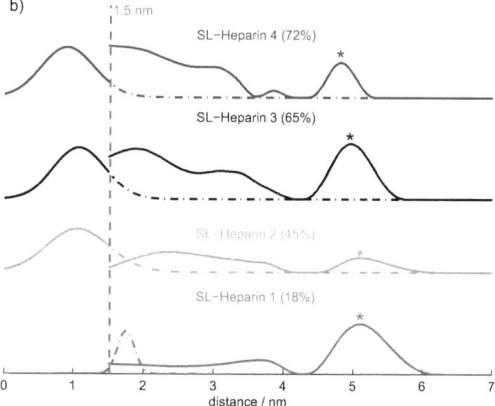

Figure 6.9.: Distance distribution of the spin-labels in the SL-heparins. Below 1.5 nm, the CW EPR distance distributions are plotted and above 1.5 nm the DEER distance distributions are plotted: SL-heparin 1 (18%, red), SL-heparin 2 (45%, green), SL-heparin 3 (65%, black), SL-heparin 4 (72%, blue). The magenta colored vertical line corresponds to the Larmor frequency of ^{1}H at 0.345 T. The distance marked with an asterisk at 5 nm might be an artifact.

6. Solid-State DNP Performance of Polarizing Agents

has a length of about 1.75 nm.[88] Due to the helical structure of spin-labeled heparin, we find not only electron spin distances along the long axis but also in radial directions[89] since two successive units have a rotation angle of about 180 degree. According to the crystallographic structure of heparin, the distances

- of the C-atoms of the carboxy-group (SL-heparin 1, 2 and 4) are
    - 0.97 nm
    - 2.66 nm and
    - 4.39 nm for opposing atoms and
    - multiples of 1.75 nm for axial atoms.
- of the N-atoms of the amino-group (SL-heparin 3) are
    - 1.08 nm,
    - 2.70 nm and
    - 4.42 nm for opposing atoms and
    - multiples of 1.75 nm for axial atoms, as well.

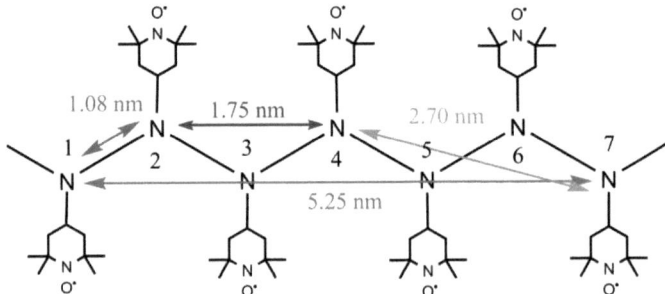

Figure 6.10.: Scheme of the heparin helix as seen from lateral view (*cf.* Figure 6.5). The scheme is not to scale but shall give a better understanding of the possible distances between the spin labels within the macromolecule. The spin-labels are attached to the amino-group (*cf.* Figure 5.3 c) and the labeling degree would correspond to 100%. The amino-groups are labeled from 1 to 7.
Magenta: distance between amino-group 1 and 2 (direct neighbor), blue: distance between 2 and 4 (next neighbor, corresponds to the crystallographic repeat unit), green: distance between 4 and 7, red: distance between 1 and 7 (corresponds to three crystallographic repeat units).

These distances were derived from the heparin structure shown in Figure 6.5[86] and are drawn schematically in Figure 6.10. Hence, the possible electron spin intervals along the axis are 1.75 nm and multiples thereof plus a Gaussian broadening. When considering the molecular size of the nitroxide radicals for spin labels on opposing positions, the intervals at the suggested distances are elongated by a few Ångstrøm (two times 0.2 − 0.6 nm).[26]

This means that with ESE we can only detect distances up to the next-neighbouring unit. Accordingly, we can estimate reliable spin label distances from ESE measurements only for SL-heparins with a labeling degree which exceeds 50% or which are labeled via a linker so that the closest approach of the spin labels lies under 2 nm. DEER, on the other hand, is the method of choice for labeling grades below 50% *and* high labeling grades.

From these assumptions and including the size of the spin labels,[26] the dominant distances between the radicals in the different SL-heparins become (Table 6.6)[i]:

1. SL-heparin 1 (18%): 4.39 to 5.25 nm
2. SL-heparin 2 (45%): 1.75 to 2.66 nm
3. SL-heparin 3 (65%): 1.08 to 1.75 nm
4. SL-heparin 4 (72%): 0.97 to 1.75 nm.

In Figure 6.8 and 6.9 we find the maximum peak in the dipolar spectra and distance distributions for the spin-labeled macromolecules, respectively:

1. SL-heparin 1 (18%): 5.03 and 5.15 nm
2. SL-heparin 2 (45%): 2.15 and 2.36 nm
3. SL-heparin 3 (65%): 1.75 and 1.08 nm
4. SL-heparin 4 (72%): 1.43 and 0.91 nm.

For example, we derive the experimental agreement of the EPR distance measurements and the cystallographic structure for SL-heparin 3 (65%): We expect that slightly more than every second carboxy-group is labeled which means that the average distance should lie slightly below 1.75 nm but clearly above 1 nm. The maximum peak is found at 1.08

---

[i]The error from DEER data occurring due to the spin labels is maximal up to 0.4 nm

Table 6.6.: Estimated distances from x-ray structure (second column, PDB: 1HPN[86]) and from EPR measurements (last column). Note: DEER underestimates short distances whereas ESE overestimates short distances.

Radical	Estimated Interval (X-ray Structure) [nm]	Simulated Pake Pattern [nm]	Maximum Peak (DEER) [nm]	Maximum Peak (ESE) [nm]
SL-heparin 1 (18%)	4.39 - 5.25	5.03	5.15	1.76
SL-heparin 2 (45%)	1.75 - 2.66	2.15	2.36	1.06
SL-heparin 3 (65%)	1.08 - 1.75	1.75	1.87	1.08
SL-heparin 4 (72%)	0.97 - 1.75	1.43	<1.5	0.91

and 1.87 (1.75) nm for ESE and DEER analysis (dipolar spectrum), respectively. Yet, we must take into account that in ESE-detected spectra distances above 2 nm are excluded. Therefore, the measured distances from ESE-detected measurements can be seen as a lower limit. Further, we find a broad range of distances in Figure 6.9 between 1.5 and 4 nm with a small peak at 3.7 nm which is confirmed by the simulated pake pattern (3.05 nm). The feature at 3.7 nm (3.05 nm) can be associated with the length of two tetrasaccharides (corresponds to two crystallographic repeat units) which is 3.5 nm long. ESE spectra are not applicable for these large distances and can be ignored. This analysis was done for all SL-heparins and we see that the average labeling degree matches the one expected excellently. So, our results correlate perfectly with the existing distances for the amino- and carboxy-groups in the heparin macromolecule (Table 6.6). It should be mentioned that we find from our EPR experiments a pronounced peak at around 5 nm for all SL-heparins pointing out that this interval might be a multiple of the disaccharide repeat unit. This is in agreement with x-ray measurements that suggest that 5.25 nm is approximately three times the length of the crystallographic repeat unit. However, this spin label distance might also be an artifact that depends on the background subtraction as is seen in the dipolar spectra in Figure 6.8.

Conclusion - Pulse EPR

Spin-labeled heparins are biocompatible and versatile macromolecules which exhibit remarkable properties such as characteristic binding activities as well as their use for methods in magnetic resonance. By means of EPR experiments, we could probe the dipolar couplings between the electron spins and hence gain insight into the distance distributions of the spin labels that are rigidly attached to the heparin backbone. The SL-heparins intrinsically feature high dipolar electron spin-electron spin coupling fre-

6. Solid-State DNP Performance of Polarizing Agents

quencies ν_{dd} due to a high labeling degree. The distance distributions of the spin labels which were experimentally obtained show good agreement with the crystallographic structure of heparin which confirm the elongated helical-like rather than coiled structure in solution. These spin-labeled heparin macromolecules could be utilized for in vivo MRI, EPRI and the polarization technique DNP.

^{1}H DNP Enhancement via Thermal Mixing

The low temperature DNP results of SL-heparin 4 (72% labeling degree) are presented in this Subsection. Here again, a DNP profile and a microwave power sweep were performed for $c = 10$, 20 and 40 mM. The DNP parameters were determined analogue to the previous Subsection dealing with TEMPOL. Appropriately, only the results of the DNP measurements and the calculations are shown. The obtained parameters extracted from the measurements shown in Figure 6.11 are summarized in Table 6.7.

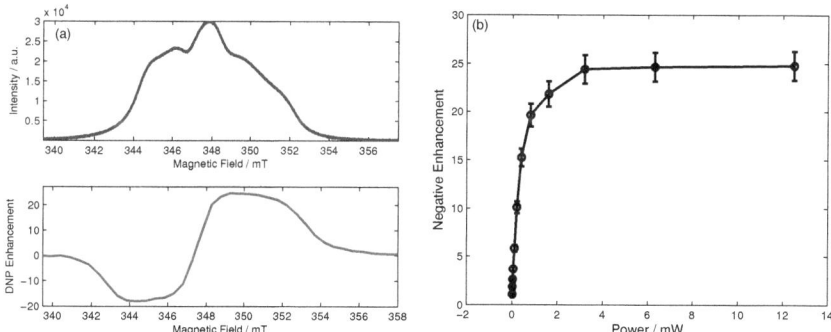

Figure 6.11.: (a) EPR spectrum and DNP profile of SL-heparin 4 (72%) and (b) its power dependence at 3493 G for $c = 40$ mM.

Table 6.7.: Summary of the important parameters of the *thermal mixing* effect for SL-heparin 4 (72%).

Concentration [mM]	ω_D [MHz]	E	$E_{\max}$	T_{1n} [ms]	W_{nD} [s^{-1}]	W [s^{-1}]
10	70.2	17.3 ± 1.4	18.0 ± 0.5	1750 ± 100	0.19	$1.6 \cdot 10^8$
20	77.2	35.2 ± 5.5	35.5 ± 2.6	1000 ± 200	0.54	$3.7 \cdot 10^8$
40	70.2	24.8 ± 1.5	26.3 ± 0.7	410 ± 20	0.83	$1.4 \cdot 10^8$

The highest enhancement was achieved for the 20 mM concentration. This sample showed the largest dipolar coupling frequency in the DNP profile and the best saturation, as well. The nuclear-spin lattice relaxation times are comparable to the ones obtained for TEMPOL.

Discussion

SL-heparin is characterized by dipolar broadened EPR spectra as seen by CW EPR and ESE-detected spectra and DEER data (*cf.* Figures 6.6 and 6.7). Due to the statistically attached spin-labels, a wide range of electronic dipole-dipole coupling frequencies is covered. Along with the well-established extraordinary enhancement of biradicals[60,63] as compared to mono-radicals originating from a dipolar coupling frequency matching the nuclear Larmor frequency, one can assume a similar effect for a spin-labeled macromolecule like SL-heparin. It is shown in the EPR results of this Section that the proton Larmor frequency is covered by SL-heparin 4 (72%). Obviously, by comparing the Tables 6.2 and 6.7 one finds that SL-heparin yields equal or less enhancement as compared to TEMPOL. Accordingly, the additional effect of biradicals is not seen for the investigated SL-heparins under the used experimental conditions. Nonetheless, at $c = 20$ mM, the enhancement of TEMPOL could be reproduced. The calculated *thermal mixing* parameters for T_{1n}, ω_D, W_{nD} and W are in the same range of magnitude as for TEMPOL. The comparison to other publications is already covered by the previous Section 6.1. The only difference to be found is the strength of the dipolar coupling frequency which is relatively independent on the concentration and subsequently elevated at low concentrations as compared to TEMPOL. The analysis of the dipolar coupling strength of the spin-labels is fully covered in the previous Subsection. Interestingly, a medium concentration ($c = 20$ mM) produces the highest enhancement which is even comparable to TEMPOL. Usually, the enhancement achieved by the *thermal mixing* effect peaks around 40 to 50 mM. As a result, at least the amount of radicals needed to achieve a desired signal enhancement can be reduced. This leads to narrower NMR line widths which are favorable for NMR measurements as the NMR line widths increase with the radical concentration. Additionally, as mentioned before, the SL-heparins are functional and biocompatible substances what is usually seen as the main drawback of polarizing agents.

Conclusion

It was proven by Hu *et al.*[60] that dipolar coupling frequencies caused by biradicals can improve the solid-state DNP. However, here it was shown that anisotropically distributed spin-labels which cover a wide range of dipolar coupling frequencies do not show a similar effect as observed for biradicals. Nonetheless, it was found that the enhancement peaks at a concentration below 40 mM and that this is comparable to the enhancement of TEMPOL. The use of less radicals leads to narrower line widths in the NMR spectrum and an increase of the nuclear spin-lattice relaxation time. Moreover, the SL-heparins are biocompatible.

6.3. Thermoresponsive Spin-Labeled Hydrogel

So far, only one procedure for radical separation for the dissolution DNP is proposed in literature in which the radicals are filtered by the use of ion-exchange columns.[13] In this work, for the first time the use of thermoresponsive, spin-labeled hydrophilic polymer networks (SL-hydrogels)[15,82] for solid-state DNP and especially dissolution DNP is introduced. Supplementary, it was already demonstrated in Section 5.3 that they allow fast and simple radical - solute separation at room temperature (*cf.* Figure 5.9). The synthesis and the sample preparation are described in Section B.3. Heating of the swollen hydrogel to the critical temperature $T_C = 63$ °C induces a reversible, fast ($\leq$ 1s) and dramatic volume decrease (≥ 500 volume%) thereby expelling water and other target molecules from the radical-bearing polymer network.[83]

Solid-State DNP Performance

In this Section, the DNP enhancement of the introduced SL-hydrogels at low temperatures is demonstrated. The low temperature measurements were only performed for SL-hydrogel-2 (*cf.* Figure B.1 in Appendix B.3) as the *thermal mixing* effect requires a high radical concentration. The concentration of ~ 1 mM for SL-hydrogel-1 is consequently far too low. The presented DNP enhancement can be seen as a proof of principle for the use of SL-hydrogels. The sample preparation for these measurements is described in Section B.3. To determine the parameters of the *thermal mixing* process, again a DNP profile and a power dependence were recorded. These measurements are visualized in Figure 6.12. The dipolar coupling frequency could be determined to 63.2 MHz which lies in between the values obtained for TEMPOL and SL-heparin 4, respectively. Saturation recovery experiments showed interestingly short T_{1n} times of ~ 60 ms. The effect of strongly shortened T_{1n} times of water in hydrogel networks was already measured at room temperature.[15,68] The best enhancement was $E \sim 4$ which is one order of magnitude lower than for TEMPOL and SL-heparin 4. The origin of this result covers the short nuclear spin-lattice relaxation time, the very low radical concentration and the limited microwave power. One can read from Figure 6.12 (b) that even at 200 mW the sample is not yet completely saturated. Unfortunately, the T_{1e} was not measured so that the saturation parameter could not be calculated.

6. Solid-State DNP Performance of Polarizing Agents

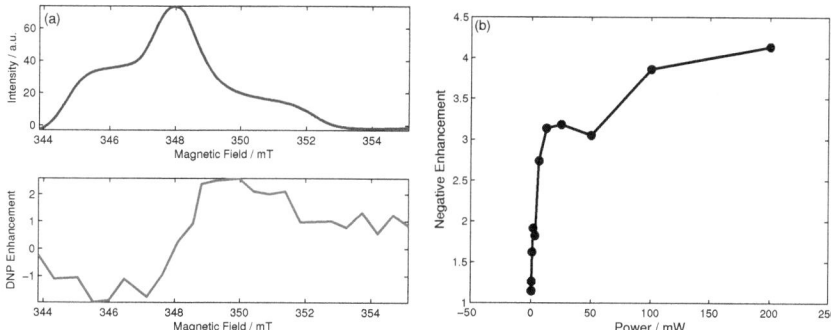

Figure 6.12.: (a) EPR spectrum and DNP profile of SL-hydrogel-2. The extrema were estimated to $B_- = 3455$ G and $B_+ = 3500$ G. (b) Microwave power dependence of the measured NMR signal enhancements for SL-hydrogel-2. The full saturation is not reached yet for 200 mW. Unfortunately, 200 mW is the maximum output power achievable with the used microwave system.

Table 6.8.: Summary of the important parameters of the *thermal mixing* effect for SL-hydrogel-2.

Concentration [mM]	ω_D [MHz]	E	T_{1n} [ms]	W_{nD} [s^{-1}]	W [s^{-1}]
~ 6.8	63.2	4.1 ± 0.2	59 ± 3	0.68	-

Discussion

The preliminary solid-state DNP results of SL-hydrogel can be seen as proof-of-principle that they are feasible to hyperpolarize nuclear spins even at X-band and relatively high temperatures (9 K). The enhancement of $E \sim 4$ could be further increased by

(i) drastically increasing the radical concentration,

(ii) applying higher microwave powers and

(iii) cooling down to ~ 1 K.

Work along these lines in a collaboration with the Grütter group at the EPFL in Lausanne is in progress. Once a sufficient degree of polarization is achieved for these thermoresponsive SL-hydrogels, the hyperpolarized samples benefit from prolonged T_{1n} times and narrow NMR lines as already discussed in Section 5.3.

Conclusion

In summary, due to the collapse of the hydrogel at elevated temperatures, the presented system is predestined for the dissolution DNP. The obvious benefits are the prolonged T_{1n} after the dissolution step and the *radical-free* and *non-toxic* solute containing the hyperpolarized biomolecules allowing for biomedical applications. These unique properties make the studied system a good polarizing agent for *in situ* and *ex situ* DNP experiments, as well. For *in situ* DNP methods the radical induced line broadening could be avoided and due to the prolonged T_{1n} time more complex 2D NMR experiments could be performed which might be beneficial for the investigation of biomolecules. Regarding the *ex situ* DNP methods the most obvious benefit lies in the prolonged lifetime of the hyperpolarization during the transport time thereby reducing the polarization loss until the utilization of the hyperpolarized molecules. Furthermore, a magnetic field of 3.5 T and $T \sim 1$ K in combination with a high microwave power should result in a considerably larger NMR signal enhancement.

6.4. Summary - Solid-State DNP

The experiments at cryogenic temperatures ($T \sim 9$ K) demonstrate the versatility and reliability of the DNP set-up. In addition, the cryogenic temperature was achieved with a closed cycle cryostat which implies no helium consumption and consequently considerably lower running costs. Under similar experimental conditions the enhancement was even higher than published by Gerfen *et al.*[59] The reasoning is the scaling of the enhancement with the magnetic field ($E \propto \frac{1}{B_0}$). The measured and calculated *thermal mixing* parameters were shown and compared to a publication by Duijvestijn *et al.*[49] where similar results are presented. Remarkably, the T_{1n} time is two orders of magnitude smaller than expected. The origin of this large difference could not be found.

The hypothetic assumption that anisotropically distributed spin-labels which cover a wide range of dipolar coupling frequencies would show a similar effect as it was observed for biradicals[60,63] could not be verified. Potentially, the additional effect caused by dipole-dipole couplings can be validated at higher magnetic fields. Nonetheless, it was found that the enhancement peaks at a concentration below 40 mM and is comparable to the enhancement of TEMPOL. Moreover, the SL-heparins are biocompatible and functional macromolecules extending their potential use as compared to TEMPOL and biradicals.

In the last Section preliminary solid-state DNP results for SL-hydrogel are shown. The enhancements were only moderate but could be increased by increasing the radical concentration, applying higher microwave power and cooling to lower temperatures. Once a sufficient polarization is achieved the benefits of the use of SL-hydrogels are the prolonged T_{1n} times and narrow NMR lines of the hyperpolarized nuclear spins.

7. Hyperpolarization of Hetero Nuclei

Nuclear magnetic resonance and related techniques have become indispensable tools with innumerable applications in physics, chemistry, biology and medicine. One of the main obstacles in NMR is its notorious lack of sensitivity which is due to the low equilibrium polarization of nuclear spins at ambient temperature. This lack becomes obvious if low γ nuclei (e.g. ^{13}C or ^{15}N) are employed for NMR spectroscopy and imaging or if small sample volumes should be investigated. However, this obstacle could be overcome by *in vitro* hyperpolarization of molecules via DNP and subsequent injection into the animal or patient of investigation as was demonstrated recently.[16,90–93] MR angiography and perfusion measurements can benefit significantly from using hyperpolarized ^{13}C containing molecules since there is no background signal from the thermally polarized ^{13}C nuclei in the tissue, and therefore the contrast-to-noise ratio approaches the signal-to-noise ratio.[91,92] Additionally, the theoretical enhancements which can be achieved for hetero nuclei are much larger than for protons as their magnetogyric ratio is lower than the one of ^{1}H spins. Another emerging field in medical applications for the use of hyperpolarized substances is molecular or real time metabolic imaging, *e.g.* to improve the diagnostic assessment of tumors or to monitor their therapy. Pathologic tissue can be differentiated from normal tissue on the basis of different metabolic pathways within the citrate cycle. As a consequence, the NMR signal intensity of lactate after injection of hyperpolarized ^{13}C-pyruvate is significantly elevated in cancer when compared to that in normal tissue.[16,17,93] One major issue of this approach is the limited lifetime of the hyperpolarized state, thereby restricting the application and detection of the hyperpolarized molecules to roughly 3 times T_{1n}. Therefore, a lot of effort is put into the hyperpolarization of biomolecules with long spin lattice relaxation times.

In this Chapter, the first direct Overhauser-type DNP experiments of ^{13}C and ^{19}F in a mobile set-up are presented. In general, the investigation of hetero nuclei can be divided into two subgroups:

(i) The molecules containing the hetero nuclei are dissolved in a solvent.

(ii) The solvent itself contains the hetero nuclei.

In the first case the DNP effect depends on the motion of the polarizing agents, the dissolved molecules and the solvent molecules which determine the direct coupling of radicals and hetero nuclei. This dependence on the correlation time was introduced in Chapter 2.3. Furthermore, the indirect polarization of the hetero nuclei mediated by the hyperpolarized solvent can play a non-negligible role. This indirect polarization which is introduced as three-spin effect in Chapter 2.3 can be in favor of or to the disadvantage of the direct DNP enhancement. In the latter case, obviously the indirect polarization depends on the properties of the solvent. If the solvent does not contain any protons the three-spin effect can be neglected. In the following Sections selected experiments with differing solvents and solutes are discussed.

7. Hyperpolarization of Hetero Nuclei

7.1. Hexafluorobenzene

Hexafluorobenzene (HFB) is a non-polar solvent and consists of six fluorine atoms attached to a benzene ring. The natural abundance of fluorine-19 is nearly 100%. The molecular structure is depicted in Figure 7.1. Additionally, the magnetogyric ratio of ^{19}F is comparable to that of ^{1}H. Therefore, it is easily detectable by NMR. Furthermore, HFB possesses a low dielectric constant ($\epsilon_r \sim 2$) which makes it an ideal solvent for DNP from the EPR point of view. Due to the low dielectric constant, one observes only vanishing absorption of the electrical component of the microwave which allows for the investigation of large sample amounts and simultaneously for high microwave power irradiation without heating effects. To investigate the DNP enhancement of ^{19}F, the radical TEMPO instead of TEMPOL is dissolved in HFB as TEMPO is not as polar as TEMPOL. The investigated concentrations of TEMPO are 2.5, 5, 10 and 20 mM. The achievable enhancement for a pure dipolar coupling is due to the lower magnetogyric ratio of fluorine slightly elevated as compared to protons: $E_{\text{theo}} \approx -350$. Additionally, HFB contains no protons so that the three-spin effect via protons as introduced in Section 2.3 can be ruled out.

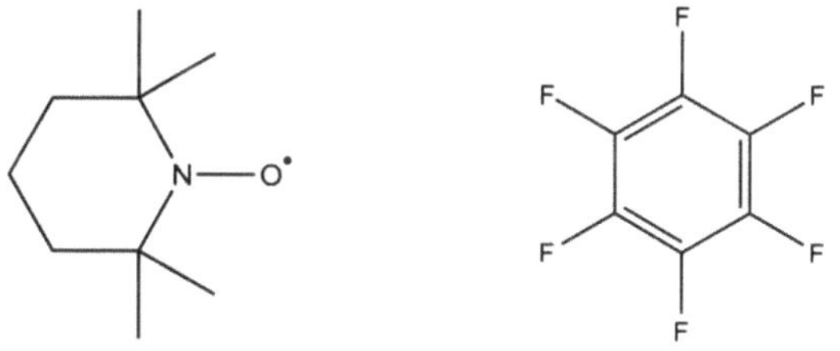

Figure 7.1.: Molecular structure of TEMPO (left) and hexafluorobenzene (right). Concerning the three-spin effect, it should be mentioned that hexafluorobenzene does not contain any protons. As a result, the three-spin term can be neglected for this radical-solvent system.

Results

CW EPR

The CW EPR spectrum of TEMPO dissolved in hexafluorobenzene for $c = 2.5$ and 20 mM is shown in Figure 7.2 on the left hand side. The CW EPR spectrum of TEMPOL

($c = 10$ mM) in water is plotted for comparison. Obviously, the EPR lines of TEMPO in HFB are broader as compared to TEMPOL in water although they exhibit similar correlation times ($\tau_c \sim 55$ ps). This can be attributed to a relatively high oxygen concentration in HFB. The air-saturated oxygen concentration in most organic solvents is approximately $1-2$ mM/l according to a publication by Hausser.[94] The dissolved oxygen molecules collide with TEMPO radicals thereby broadening the EPR lines analogous to HSE. Further, this coupling of unpaired nitroxide electrons and the paramagnetic oxygen molecules is dominant compared to all other interactions[94] like electron-nuclear spin coupling. The exchange rate, induced by dissolved oxygen, is determined qualitatively fitting a simulated spectrum to the experimental one. The fitting results give an exchange rate of a few GHz for TEMPO ($c = 10$ mM) in HFB as compared to an exchange rate of a few MHz for TEMPOL ($c = 10$ mM) in water as shown exemplary in Figure 7.2. The high exchange rate results in broad EPR lines and short electronic spin-lattice and spin-spin relaxation times.

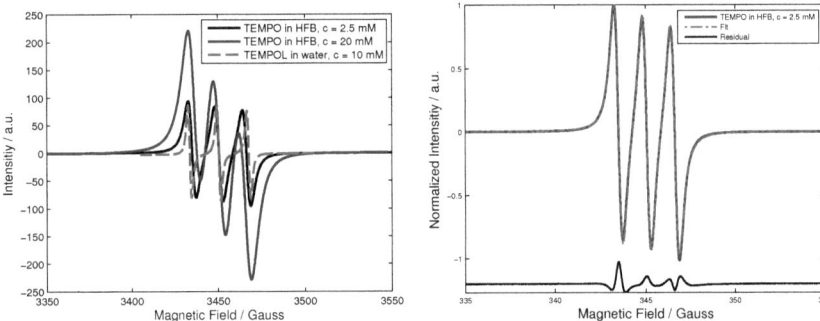

Figure 7.2.: Left: Comparison of experimental CW EPR spectra of TEMPO in hexafluorobenzene at two selected radical concentrations to TEMPOL in water. Black: TEMPO (HFB), $c = 2.5$ mM, Blue: TEMPO (HFB), $c = 20$ mM, Red: TEMPOL (H$_2$O), $c = 10$ mM. Right: The fit of the experimental spectrum utilizing the function "esfit" from the Easyspin software package yields that the CW EPR lines are broadened due to exchange interactions (exchange rate ~ 4 GHz). The correlation time ($\tau_c \sim 55$ ps) is similar to that of free TEMPOL in water ($\tau_c \sim 30$ ps).

^{19}F DNP Performance

The leakage and the coupling factor are determined via nuclear spin-lattice relaxation experiments as introduced before. The leakage factor is slightly lower than the values for free TEMPOL in water (cf. Tables 5.2 and 7.1). The coupling factor shows a strong deviation between the different TEMPO concentrations as the T_{1n} measurements at the operating field featured large errors. Therefore, the determined coupling factors vary from 0.25 up to 0.45. The values of f and ξ for all concentrations are summarized in Table 7.1. The saturation is calculated using Equation 2.40 and the parameters given in Table 7.1. For the given values of f and ξ, the saturation is only $1-2\%$ due to the low achieved signal enhancement.

Table 7.1.: DNP parameters describing TEMPO in hexafluorobenzene.

TEMPO	E	E_{max}	f	ξ	s_{eff}
2.5 mM	-4.1 ± 0.1	-4.5 ± 0.2	0.29 ± 0.12	0.25 ± 0.28	0.015 ± 0.031
5 mM	-8.1 ± 0.2	-10.8 ± 0.4	0.53 ± 0.09	0.42 ± 0.12	0.013 ± 0.017
10 mM	-12.1 ± 0.2	-19.9 ± 0.9	0.58 ± 0.08	0.28 ± 0.11	0.012 ± 0.008
20 mM	-11.3 ± 0.1	-19.2 ± 0.2	0.81 ± 0.04	0.45 ± 0.09	0.010 ± 0.006

EPR Frequency Dependence of the ^{19}F DNP Enhancement

A plot of the DNP enhancement of TEMPO (10 mM, HFB) as a function of the magnetic field reflects the measured EPR spectrum. This is visualized in Figure 7.3 where the DNP enhancement follows the integrated CW EPR spectrum. The center line ($\Delta m_I = 0$) clearly shows the highest enhancement ($E = -5$) as expected. The high field line ($\Delta m_I = -1$) exhibits the lowest enhancement factor ($E = -4$). In between the EPR lines the enhancement does not drop below $E = -3$. The effect that the enhancement does not drop to zero in between the hyperfine lines reflects the short electronic relaxation times of TEMPO in HFB, especially T_{2e}.

Discussion

The negatively enhanced signals imply that the dipolar dominates over the scalar interaction although the radical-nucleus interaction can contain a scalar part, too.[95] The

7. Hyperpolarization of Hetero Nuclei

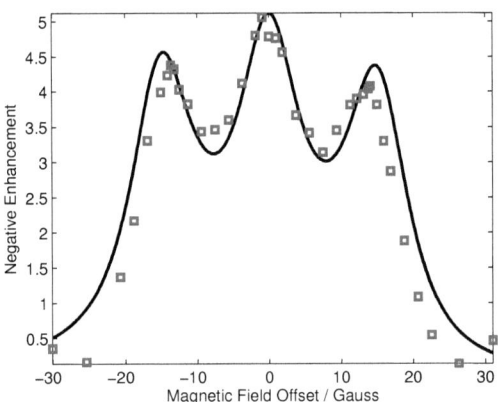

Figure 7.3.: Comparison of experimental CW EPR spectrum (first integral, 10 mM TEMPO in HFB) with ^{1}H DNP enhancement curve at room temperature. The solid line depicts the integrated EPR spectrum normalized to the maximum enhancement and squares the absolute DNP enhancement as a function of the magnetic field offset (microwave power = 4 W). Between the hyperfine lines the enhancement does not drop below 60% of the maximum enhancement but follows the EPR line. This effect shows the significant overlap of the hyperfine lines when one irradiates off-resonant.

achieved DNP enhancements lie far below the theoretical maximum of $E_\text{theo} \approx -350$. The leakage and the coupling factor show similar values as for free TEMPOL in water, though. Consequently, the correlation times of the radicals in both solvents should be similar. The EPR result confirms that the motion of the nitroxide radicals in HFB is almost unchanged as the fit gives for TEMPO in HFB $\tau_c = 55$ ps which is close to $\tau_c = 30$ ps for TEMPOL in water (cf. Figure 7.2). Contrary to the large f and ξ value, the effective saturation s_eff is only $1-2\%$ and is clearly the limiting factor for ^{19}F DNP enhancement in organic solvents. The reason for the almost vanishing saturation can be found in the high oxygen concentration in air-saturated organic solvents as mentioned before. The interaction of TEMPO and oxygen radicals is extremely effective as was already observed by Hausser et al.[96] First, this leads to a reduced T_{1e} and T_{2e} relaxation time of the radical. Subsequently, the saturation factor is diminished (cf. Equation 2.43) as broad lines are more difficult to saturate. Short relaxation times are also the reason for the broad DNP profile in Figure 7.3. It was introduced in Chapter 2.3 that short relaxation times lead to a non-vanishing enhancement in between EPR hyperfines lines. Second, the interaction between unlike paramagnetic agents does not result in an enhanced saturation like the encounter of like radicals (cf. Chapter 5.2 and references[31,41,94,96]). So, this interaction can be seen as a leakage in the considered system which severely lowers the achievable saturation. The degree of loss can only be estimated from the ratio of the spin exchange rates of TEMPOL in water and TEMPO in HFB. The ratio is approximately 10^{-3} which confirms the very high efficiency of this relaxation pathway. This low ratio comes as a surprise as the radical concentration exceeds the one of oxygen by up to one order of magnitude.

Additionally, the radical-oxygen interaction provides a contribution to the spin-lattice relaxation time of the fluorine nuclei and diminishes the actual leakage factor. Thereupon, the coupling factor is also influenced as it is measured by means of spin-lattice relaxation experiments. This leads to an overestimation of the coupling factor which further reduces the calculated saturation. Finally, Hausser et al.[96] observed a decrease of the proton signal enhancement of the order of two magnitudes when they measured with organic, air-saturated solvents as compared to oxygen-free samples. Therefore, a de-oxygenation should result in considerably higher NMR signal enhancements in organic solvents but could not be verified due to time limitations.

Conclusion

For the first time fluorine spins were hyperpolarized in our research group by means of the DNP technique. The resulting enhancements showed slightly enhanced and inverted NMR signals which proves that the interaction between electron spins and fluorine nuclei is mainly of dipolar origin. The leakage and the coupling factor showed values comparable to free TEMPOL. These values are overestimated as dissolved oxygen molecules also contribute to the spin-lattice relaxation of ^{19}F which is not taken into account. As a result the small measured enhancements stem from the saturation factor which is only $1-2\%$ due to the high efficiency of a relaxation pathway which is not in favor of the DNP effect. Ergo, from the EPR point of view hexafluorobenzene is a perfect solvent for DNP due to its low dielectric constant, but it is a bad solvent in regards of the achievable DNP enhancements due to dissolved oxygen. Further, DNP measurements in organic solvents should be performed under exclusion of oxygen.

7.2. DNP of a Dissolved ^{19}F Containing Molecule

In a collaboration with Dr Kristofer Thurecht (Center for Advanced Imaging, University of Queensland, St Lucia, Queensland, Australia) the feasibility of measuring ^{19}F spins of molecules dissolved in water was tested. The used molecule (Figure 7.4 and Appendix B.5) is a statistical, hyperbranched polymer. On average, the hyperbranched polymers were 8-arm stars with a weight of 22 kDa. This molecule was originally synthesized for the use as ^{19}F-marker in MRI to follow its metabolic pathway. The hyperpolarization of the fluorine by DNP can lead to a significant sensitivity increase for this molecule.

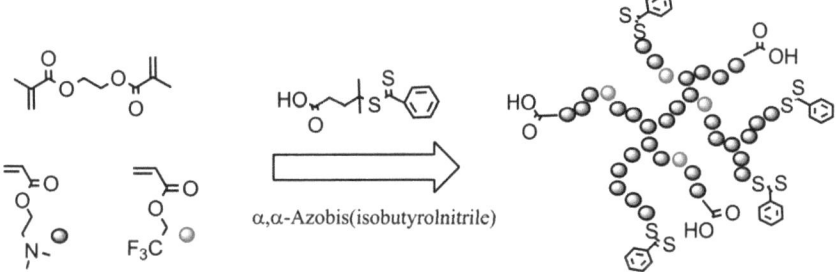

Figure 7.4.: Molecular structure of the described hyperbranched polymer. The acronym for this molecule is KJT.

Due to sensitivity reasons, the measurements were performed in an electromagnet. The T_{1n} time of ^{19}F in this molecule at X-band and room temperature was measured via an inversion recovery experiment with cpmg-detection to be 140 ± 10 ms. Accordingly, a microwave irradiation time of one second was sufficient. The highest enhancement was observed for an irradiation time of 3 s and a microwave power of 4 W which is sufficient to completely saturate an aqueous 20 mM TEMPOL sample. The corresponding measurement is shown in Figure 7.5 where the reference spectrum with 1000 accumulated scans is plotted for comparison, as well. The computation yielded an absolute enhancement of $E = -37.3 \pm 0.6$ which equals an acquisition time saving of ~ 1400. Indeed, the DNP-enhanced spectrum (single shot) showed a slightly better signal-to-noise ratio than the reference spectrum with 1000 scans. Interestingly, the fluorine atoms are located at the inner parts of the KJT molecule which hinders the direct radical-^{19}F contact. The steric hindrance significantly lowers the coupling factor which explains the reduced enhancement as compared to water protons. In addition, the three-spin effect due to the water protons further reduces the possible DNP enhancement. To eliminate the three-spin effect, one can either saturate the proton transition or dissolve the KJT molecule in D_2O as explained in Section 2.3. Nevertheless, these preliminary DNP results of a large macromolecule are promising steps to the hyperpolarization of proteins.

Unfortunately, a power dependence for this sample could not be measured and the concentration of KJT of the used sample is unknown yet. As a result, the DNP enhancement of a dissolved ^{19}F containing molecule presented here can be seen as a proof of principle for the direct hyperpolarization of a large polymeric molecule. It demonstrates the sensitivity which can be achieved at 0.345 T by utilizing DNP.

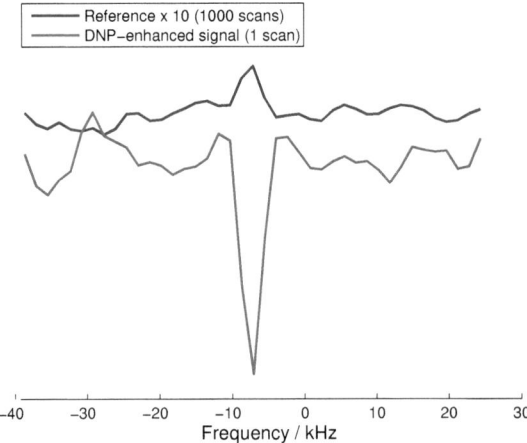

Figure 7.5.: Comparison of the reference spectrum (20 mM TEMPOL, 1000 scans) and the DNP-enhanced single-shot ^{19}F signal (20 mM TEMPOL, 1 scan). The reference spectrum was multiplied by a factor of ten for better visualization.

7.3. ^{13}C-Enriched Urea

The main disadvantages of organic solvents are their high amount of dissolved oxygen and their non-compatibility with proteins and biomolecules. Usually, organic solvents denature proteins which prevents their use in the investigation of proteins via DNP. For this reason, the studied biomolecule urea is dissolved in water. The concentration of dissolved urea in the aqueous samples was chosen as high as possible in order to provide as many ^{13}C spins as possible. The actual concentration of the ^{13}C-labeled urea was $c = 10$ M. TEMPOL was used as polarizing agent as it still provides the highest NMR signal enhancement observed for the Overhauser-type DNP. The selected concentrations of TEMPOL were 10, 20, and 40 mM. The concentration range was selected relatively high in order to shorten the ^{13}C nuclear spin-lattice relaxation time. The results presented in this Section were obtained with the mobile X-band system.

^{13}C DNP Enhancement in the Mobile Set-up

The NMR measurements of this Subsection were exclusively recorded in the cpmg-detection mode in order to maximize the signal-to-noise ratio. The DNP measurements were performed with capillaries which contained a sample volume of $V \sim 2.5$ μl. The reference measurements (no microwave irradiation) of capillaries were not accessible due to the low magnetogyric ratio of ^{13}C and the resulting NMR insensitivity of small sample amounts. For this reason, the reference experiments had to be measured in 3 mm tubes which contained a sample volume of $V \sim 25$ μl. Consequently, the DNP enhancement was estimated by multiplying the calculated enhancement by a factor of ten, which corresponds to the volume ratio.

Results

The feasibility of doing single-shot DNP experiments in a mobile X-band system is visualized in Figure 7.6. Here, the enhancement was estimated to $E = -324 \pm 128$ which corresponds to a reduction of acquisition time of $> 10^5$. The large error estimates stem from the low signal-to-noise ratio of the ^{13}C NMR signal at 0.35 T. Furthermore, the DNP enhancement was measured in dependence of the microwave power in order to determine the maximum achievable enhancement in our mobile system. For this experiment, the irradiation time was kept constant via a LabView-based remote control and the microwave was allowed to vary. The results of this measurement are plotted in

Figure 7.7.

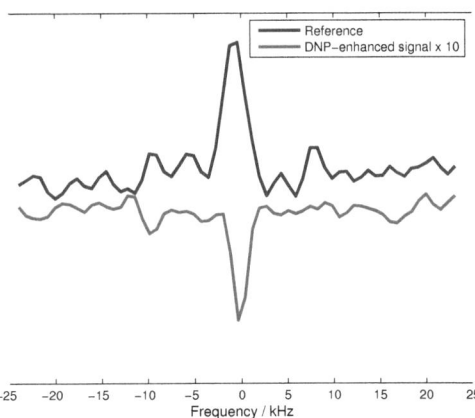

Figure 7.6.: Comparison of the ^{13}C urea reference spectrum (40 mM TEMPOL, 500 scans, 25 μl sample volume) and the DNP-enhanced single-shot ^{13}C signal (40 mM TEMPOL, 1 scan, 2.5 μl sample volume). The DNP-enhanced spectrum was multiplied by a factor of ten to account for the difference in the sample volume. Both experiments were recorded in the cpmg-detection mode. The irradiation time and the microwave power were 5 s and 4 W, respectively. The estimated enhancement is $E = -324 \pm 128$. This corresponds to a saving of acquisition time of $> 10^5$. The large error estimates stem from the low signal-to-noise ratio of the ^{13}C NMR signal.

Thus, the theoretical maximum enhancement reaches a value up to $E_{\text{max}} = -574 \pm 206$ if one could saturate the sample completely. Figure 7.7 (a) shows only the measurable DNP-enhanced NMR signals because below 0.5 W the signal was not detectable any more.

Due to the relatively long T_{1n} times of ^{13}C (several seconds), not only a power sweep (at constant irradiation time) but also varying the irradiation time (at constant microwave power) is feasible. This measurement was performed via the LabView-based remote control which allows for the adjustment of very accurate irradiation times. The accuracy of the irradiation times is essential for the reliability of the DNP build-up experiments. For example, Figure 7.8 shows the obtained area of the DNP-enhanced signal in dependence of the microwave irradiation time for ^{13}C urea ($c = 10$ M) and TEMPOL ($c = 20$ mM) in

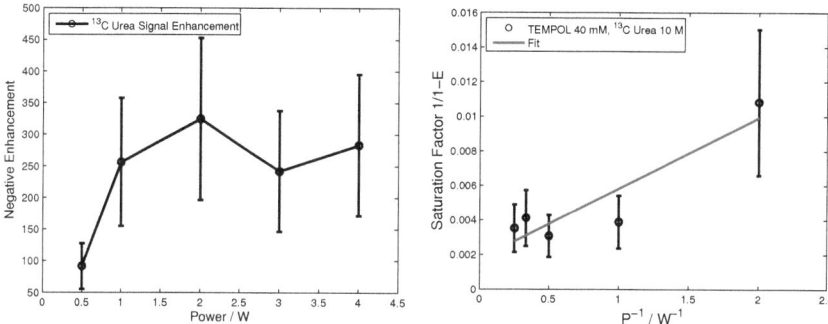

Figure 7.7.: (a) DNP enhancement of ^{13}C in dependence of the irradiated microwave power. Below 0.5 W the NMR signal could not be detected any more. (b) Plot for the determination of the maximum achievable enhancement in our mobile system. The maximum achievable enhancement with full saturation is $E_\text{max} = -574 \pm 206$.

water. Theoretically, by measuring the so-called DNP build-up time one can determine the T_{1n} time of the hyperpolarized nucleus as the enhancement increases exponentially with the spin-lattice relaxation time. This curve increases exponentially and can be fitted with the function

$$A = B \cdot \left(1 - \exp\left(\frac{t}{Q}\right)\right) , \tag{7.1}$$

in which A is the calculated area of the signal, B an unimportant fitting parameter, t the irradiation time and Q the fitted DNP build-up time constant. Ergo, Q can be identified as the nuclear spin-lattice relaxation time T_{1n} of the hyperpolarized nucleus. The irradiation time and microwave power dependence were measured for all concentrations. The summary of these results is shown in Table 7.2.

With the calculated T_{1n} times the crucial factors f, ξ and s_eff of the Overhauser Equation can be obtained. This was done as described in Section 5.1. The obtained parameters are also listed in Table 7.2.

7. Hyperpolarization of Hetero Nuclei

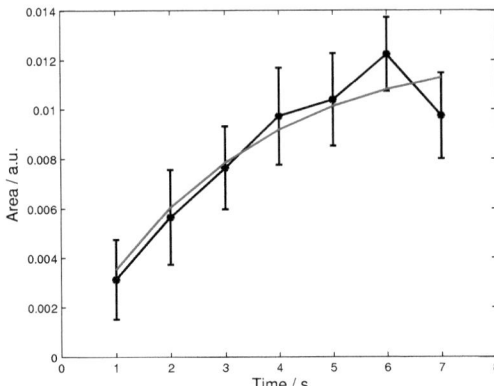

Figure 7.8.: Plot of the DNP enhancement in dependence of the microwave irradiation time at 4 W. The enhancement increases exponentially and can be fitted via Equation 7.1. The determined build-up time corresponds to the nuclear spin-lattice relaxation time of the hyperpolarized nucleus. For ^{13}C of urea in an aqueous solution of 20 mM TEMPOL, the fit yielded $T_{1n} \sim 3.1$ s. The large error estimates stem from the low signal-to-noise ratio of the ^{13}C NMR signal.

Table 7.2.: DNP parameters of the investigated samples (TEMPOL and urea dissolved in water).

TEMPOL	E	E_{max}	T_{1n} [s]	f	ξ	s_{eff}	$T_{1n}{}^d$ [s]
10 mM	-71.4 ± 22.6^a	-71.2 ± 15.2	6.4 ± 1.0	0.82 ± 0.03	0.11 ± 0.23	0.17 ± 0.10	6.12 ± 0.21
20 mM	-257 ± 43^b	$-391\pm$ NaN	3.1 ± 0.6	0.90 ± 0.02	0.19 ± 0.41	0.55 ± 0.28	2.79 ± 0.07^e
40 mM	-324 ± 128^c	-574 ± 206	1.6 ± 0.3	0.95 ± 0.01	0.30 ± 0.19	0.66 ± 0.40	1.58 ± 0.10

[a] mw irradiation time: 2 s; [b] mw irradiation time: 4 s; [c] mw irradiation time: 5 s; [d] determined at an electromagnet by K. Gruss[97] without utilizing DNP; [e] $c = 25$ mM

Discussion

The feasibility of performing single-shot ^{13}C experiments in a mobile X-band set-up could be shown by two measurements. The first measurement is a simple microwave power dependence which allows for the analysis of the maximum achievable enhancement under the assumption of full saturation. The highest measured enhancement in this experiment was $E = -324$ which equals a reduction of acquisition time of $> 10^5$. Upon full saturation, this enhancement could be increased to $E_{\max} = -574$. These are the largest ever reported enhancements for ^{13}C at X-band. Additionally, these are the first ^{13}C DNP-enhanced signals in a mobile NMR system at 0.35 T. The measured results can only be compared with a publication[47] obtained in an electromagnet. Otherwise, the experimental conditions were the same. The comparison ($c = 20$ mM) yields that our experimental set-up is comparable in the DNP performance. Lingwood et al.[47] measured $E = -265$ ($E_{max} = -455$) which is close to the value of $E = -257$ ($E_{max} = -391$). It is noteworthy that in this work the ^{13}C measurements were performed in a Halbach magnet which clearly shows a higher inhomogeneity which leads to the conclusion that either our probehead is better or that the EPR line width is not as severely broadened as expected. The comparison to the results obtained by our group in an electromagnet set-up[97] show similar results as published by Lingwood et al.[47] and are consequently not further discussed.

Another important aspect is the microwave irradiation time. On the one hand, a long irradiation time can lead to an evaporation of the sample. On the other hand, for a good DNP performance the sample should be irradiated for at least 3 times T_{1n}. Obviously, this is fulfilled for the concentration of 40 mM. For the concentrations 10 and 20 mM, this condition is not fulfilled any more. Especially for $c = 10$ mM, the irradiation time is only one third of the corresponding T_{1n} time resulting in a comparatively low enhancement of $E = -71$. For $c = 20$ mM, this effect is not as pronounced as the microwave irradiation time was longer than T_{1n}.

The last point concerns the three-spin effect which is explained in Section 2.3. Due to the dipolar coupling of the electron and the carbon spins, the three-spin term reduces the achievable enhancement (cf. Section 2.3). For high radical concentrations, the three-spin term approaches zero, which in turn increases the achievable enhancement. This reasoning is an additional explanation for the origin of the relatively low enhancement of the 10 mM sample. The three-spin term cannot only be suppressed by choosing high radical concentrations but also by using deuterated solvents which minimizes the coupling factor ξ_2^1, the leakage factor f_2^1 and the enhancement E_{I_1} of the three-spin term.[47]

The second measurement is a simple irradiation time dependence which allows for the computation of the T_{1n} time of the hyperpolarized nucleus as described in the previous Subsection. The determined T_{1n} times utilizing a DNP build-up curve show a perfect agreement with the values measured without DNP as can be seen from Table 7.2.

Especially the second measurement presents the sensitivity enhancement gained utilizing DNP. Without DNP the measurement of a ^{13}C T_{1n} time at such a low field takes several hours up to days. Upon microwave irradiation the same results are obtained within minutes along with a better signal-to-noise ratio. Last, the reliability of the mobile system could be shown by comparison to results obtained in an electromagnet and from another publication.[47] This clearly shows the potential of this mobile set-up for future applications.

7.4. Summary - Hyperpolarization of Hetero Nuclei

The feasibility of performing NMR experiments of hetero nuclei at X-band field strengths with a mobile set-up could be shown for ^{19}F and ^{13}C. The ^{19}F DNP experiments of the radical-solvent system TEMPO-HFB manifest the importance of dissolved oxygen in organic solvents. Without degassing the saturation reached only values of $1-2\%$. Upon degassing the saturation and thereby the enhancement should increase significantly. Another disadvantage is the incompatibility of biomolecules with organic solvents which hampers their applicability. The advantage of organic solvents is their low dielectric constant which allows for long microwave irradiation times, high microwave powers and the use of large sample amounts. Moreover, the nuclei of a ^{19}F containing molecule dissolved in water were hyperpolarized. As a result, by using DNP the NMR signal is seen in a single-shot experiment. Thus, the feasibility of detecting hetero nuclei of dissolved macromolecules at 0.345 T is verified. The promising enhancement of a macromolecule despite the steric hindrance can be seen as a further step towards the hyperpolarization of proteins at physiological temperatures.

It could be shown that in a mobile DNP polarizer similar ^{13}C enhancements are possible as in an electromagnet. The sample with the highest TEMPOL concentration clearly showed the best DNP performance owed to the shortened T_{1n} time and the reduced three-spin term. The achieved enhancement of $E = -324$ equals a saving of acquisition time of $> 10^5$. Along with this enhancement, the ^{13}C DNP measurements clearly prove the possibility of a fast and reliable NMR detection and T_{1n} determination of biomolecules by using DNP. This circumstance shows the potential of DNP, especially at X-band, when it comes to the polarization of nuclei with a very low magnetogyric ratio which should result in many new applications.

8. Conclusion

In this work, different hardware aspects and polarizing agents for dynamic nuclear polarization were studied. While the first part (Chapter 3 and 4) of this thesis mainly examined technical improvements, the last part (Chapter 5, 6 and 7) was focused on the design and the performance of new polarizing agents. The work of this thesis is summarized schematically in Figure 8.1 in which the topics of the Chapters are depicted with an expressive picture and plot, respectively.

Technical Developments

New technical components for the complete development of a mobile DNP polarizer were presented (yellow-colored part in Figure 8.1). This comprises a new Halbach magnet with a magnetic field strength up to 0.46 T and a homogeneity of $\sim$ 33 ppm. Second, the automation of the DNP experiments with well-defined trigger signals was implemented using LabVIEW. These two improvements are essential for the detection of hetero nuclei in a mobile set-up as presented in Chapter 7. Last, the construction of two probeheads were presented in which large sample volumes can be polarized. In addition, one of the probeheads possesses an integrated triple resonance which will allow for more complex NMR sequences. Nevertheless, the ENDOR probehead still shows the highest absolute DNP performance for capillaries and was used for the characterization of new polarizing agents.

Polarizing Agents for Overhauser-Type DNP

In DNP-enhanced samples the existence of radicals is always an issue. First of all, most radicals are toxic and have to be removed from the sample priori to medical applications. Furthermore, radicals lead to a shortening of the nuclear spin-lattice relaxation time which severely reduces the time frame for the utilization of a hyperpolarized sample. In addition, the radicals broaden the NMR lines. Therefore, a biocompatible spin-labeled macromolecule (SL-heparin) and a spin-labeled hydrogel network (SL-hydrogel) were investigated with respect to their DNP performance (green part of Figure 8.1).

The SL-heparins showed surprisingly high DNP enhancements which are similar to TEMPOL. In the low concentration limit their achieved NMR signal enhancements even exceed those of TEMPOL. The short T_{2e} time, which is caused by the strong residual dipole-dipole couplings, could be identified as the origin of this effect. Hence, we propose that one can optimize the use of biological systems for DNP at physiological temperatures for which only small sample amounts are available by using anisotropically distributed spin-labels on *e.g.* biological macromolecules.

Another method to overcome the occurrence of radicals in the hyperpolarized sample is the spin-labeling of radicals to a matrix. In this thesis, as substrate a thermoresponsive hydrogel was chosen. The thermoresponsitivity of the spin-labeled hydrogel predestines it as polarizing agent for DNP. Due to the collapse of the hydrogel at elevated temperatures, the presented system is characterized by a reduced nuclear spin-lattice relaxation time during the microwave irradiation and an instantaneously prolonged nuclear spin-lattice relaxation time afterwards and the benefit of a *radical-free* polarized sample.

The next step to be done in the design of sophisticated polarizing agents is the combination of the properties of SL-heparin and SL-hydrogel, which means *radical-free* hyperpolarized samples with a similar signal enhancement as obtained with free TEMPOL radicals.

Solid-State DNP of Polarizing Agents

The experiments at cryogenic temperatures ($T \sim 9$ K) demonstrate the versatility and reliability of the DNP set-up. Under similar experimental conditions the enhancement was even higher than published by Gerfen *et al.*[59] The measured and calculated *thermal mixing* parameters were shown and compared to a publication by Duijvestijn *et al.*[49] in which similar results are presented. Remarkably, the nuclear spin-lattice relaxation time is two orders of magnitude smaller than expected. The origin of this large difference could not be found.

The hypothetic assumption that anisotropically distributed spin-labels which cover a wide range of dipolar coupling frequencies would show a similar effect as it was observed for biradicals[60,63] could not be verified. Potentially, the additional effect caused by dipole-dipole couplings can be validated at higher magnetic fields. Nonetheless, it was found that the enhancement peaks at a concentration below 40 mM and is comparable to the enhancement of TEMPOL. Moreover, the SL-heparins are biocompatible

8. Conclusion

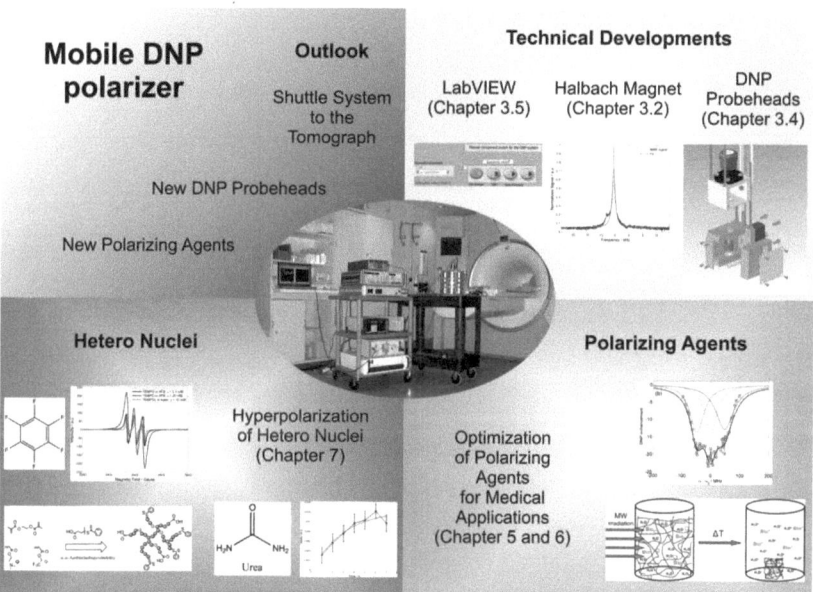

Figure 8.1.: Schematic summary of the experimental investigations in this work. The technical aspects realized in this thesis are colored in yellow. The green and blue surrounded parts comprise the development of new polarizing agents and first DNP results of hetero nuclei in the mobile set-up, respectively. The purple area shows the outlook of the mobile DNP polarizer.

and functional macromolecules extending their potential use as compared to TEMPOL and biradicals.

Preliminary solid-state DNP results for SL-hydrogel show moderate enhancement factors but could be increased by increasing the radical concentration, applying higher microwave power and cooling down to lower temperatures $T \sim 1$ K. Once a sufficient polarization is achieved the benefits of the use of SL-hydrogels are the prolonged nuclear spin-lattice relaxation time and narrow NMR lines of the hyperpolarized nuclear spins.

Hyperpolarization of Hetero Nuclei

The feasibility of performing NMR experiments on hetero nuclei at X-band field strengths could be shown for ^{19}F and ^{13}C. The DNP experiments of the radical-solvent system TEMPO-HFB show the importance of dissolved oxygen in organic solvents. Without degassing the saturation reached only a few percent. Moreover, the nuclei of a ^{19}F containing molecule dissolved in water were hyperpolarized. As a result, by using DNP the NMR signal is seen in a single-shot experiment (depicted in Figure 8.1 in blue).

Beyond, it could be shown that in a mobile DNP polarizer similar ^{13}C enhancements are possible as in an electromagnet. The sample with the highest TEMPOL concentration clearly showed the best DNP performance due to the shortened nuclear spin-lattice relaxation time and the reduced three-spin term. The achieved enhancement reduces the acquisition time by five orders of magnitude. Along with this enhancement, the ^{13}C DNP measurements clearly prove the possibility of a fast and reliable NMR detection and T_{1n} determination of biomolecules. This circumstance shows the potential of DNP, especially at X-band, when it comes to the polarization of nuclei with a very low magnetogyric ratio which should result in many new applications.

Thus, the work presented in this thesis covers both hardware aspects of the mobile DNP polarizer and molecular design of polarizing agents for DNP, hereby combining different fields of research. These new components and designs could be the foundation to open up the application of the mobile DNP polarizer for medical applications.

Outlook

The prospective use of the mobile polarizer for medical applications requires the implementation of a fast and reliable flow system. This topic will be the work of a new diploma student in our group who started in August 2010. Another improvement of the set-up can be achieved by shimming the Halbach magnet as proposed in Section 3.3. Another issue is the implementation of the triple resonant probehead which allows for more complex NMR sequences, thereby further increasing NMR signals, especially of carbon spins. This will be covered by a new PhD student who is going to start his PhD in September 2010. The last point concerns the molecular design of new polarizing agents. The design of sophisticated polarizing agents requires the combination of the properties of SL-heparin and SL-hydrogel, which means *radical-free* hyperpolarized samples with a similar signal enhancement as obtained with free TEMPOL radicals.

A. Appendix - Methods

A.1. Determination of Unknown Radical Concentrations

A plot of the double integral of the CW EPR spectra versus the concentration can be used to calibrate the conversion efficiency of the EPR spectrometer. The area of the integral of the CW EPR spectra of TEMPOL versus its concentration is plotted in Figure A.1. The fit yields a slope of (40800 ± 1249) a.u./mM. This value is used throughout this work when unknown radical concentrations had to be determined.

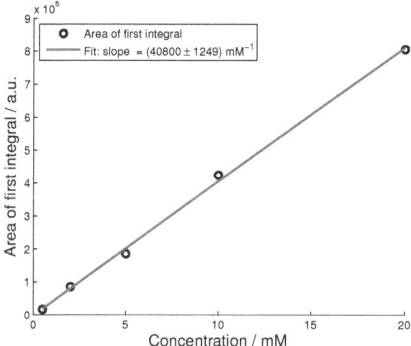

Figure A.1.: Plot of the area of the first integral of the CW EPR spectra of TEMPOL versus the radical concentration. The slope of the linear fit is used for the determination of the radical concentration of unknown samples.

A.2. CW EPR Line Shape Analysis

Determination of Electron Spin-Spin Relaxation Times

For the evaluation of the T_{2e} times of the SL-heparins, we analyzed the measurements of CW EPR spectra with a custom-written Matlab program, which utilizes the Easyspin software package for EPR[73] as described in Appendix A.2. This program accounts for the broad CW EPR lines due to the reduced mobility and the exchange interactions of the electron spins by multiplying a line shape that only reflects reduced mobility with a stretched exponential function in the time domain.[98] The uncoupled line shape was calculated with the function "chili" from the Easyspin software package. The used stretched exponential function in the time domain is also called Kohlrausch function and has the form

$$f(\tau,\beta) = exp\left[-\left(\frac{t}{\tau}\right)^{\beta}\right] \quad, \tag{A.1}$$

in which τ can be interpreted as a relaxation time T_{2e}. Subsequently, the product of uncoupled line shape and the stretched exponential function was Fourier-transformed resulting in an exchange-broadened spectrum. The best fit was found by minimizing the sum squared residual between the broadened spectrum and the experimental spectrum. Subsequently, the T_{2e} time was calculated via

$$\langle T_{2e}\rangle = \frac{\tau}{\beta}\Gamma\left(\frac{1}{\beta}\right) \quad, \tag{A.2}$$

in which Γ denotes the gamma function. For example, the measured and simulated spectra of TEMPOL ($c = 0.5$ mM) with the corresponding residual spectrum are shown in Figure A.2.

A.3. ESE-Detected Line Shape Analysis

For the quantitative analysis of the residual dipolar coupling strengths of spin-labels, the measurements of ESE spectra were analyzed with a custom-written Matlab program which utilizes the Easyspin software package for EPR.[73] This program accounts for the EPR powder spectrum and the residual dipolar couplings of the electron spins by convolving a line shape that only reflects non-broadened powder line shape with a dipolar broadening function. The uncoupled line shape was calculated with the function "pepper" from the Easyspin software package. The dipolar spectrum which represents a Gaussian distribution of distances consists of a sum of Pake patterns which were calculated with a home-written program for distances ranging from 0.1 to 5 nm and angles

A. Appendix - Methods

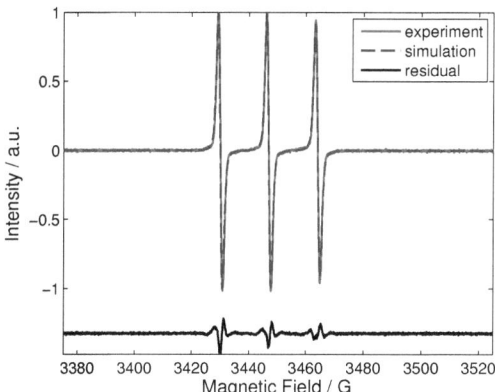

Figure A.2.: CW EPR spectrum of TEMPOL ($c = 0.5$ mM) together with its best fit according to a convolution approach explained in the text (convolved spectrum = uncoupled line shape $\otimes$ dipolar spectrum).

in the interval $[0, \frac{\pi}{2}]$. The uncoupled line shape and the dipolar spectrum were convolved by multiplying their inverse Fourier-transforms. Subsequently, this product was Fourier-transformed resulting in a dipolar broadened spectrum. The best fit was found by minimizing the sum squared residual between the dipolar broadened spectrum and the experimental spectrum.

A.4. DEER Analysis

The fitting of the DEER data was done with the Matlab-based program DEERAnalysis2008[27] which overcomes the mathematically ill-posed problem of receiving distances and distributions from the DEER time domain data by Tikhonov regularization.[24,26]

A.5. Electron spin-lattice determination at room temperature

The T_{1e} time of TEMPOL was determined via an inversion recovery experiment with FID detection. Unfortunately, the inversion recovery experiment could not be performed in an automated way via a pulse program due to a large background signal. To account for

A. Appendix - Methods

the background signal, for each delay time a FID on-resonant and a 100 G off-resonant was recorded. The background-free FID was obtained by subtracting the off-resonant from the on-resonant signal (*cf.* left hand side of Figure A.3). Subsequently, the area of the magnitude of the background-free FID was plotted versus the delay time which manifests an exponential increase (*cf.* right hand side of Figure A.3). The subsequent fit to this curve resulted in a T_{1e} time of 520 ns. The T_{1e} time was measured only for one concentration as T_{1e} does not depend on the radical concentration. Türke *et al.* measured for ^{15}N-TEMPOL a T_{1e} time of 350 ns which taking into account the differences between ^{14}N and ^{15}N is in agreement to the calculated value of 520 ns.

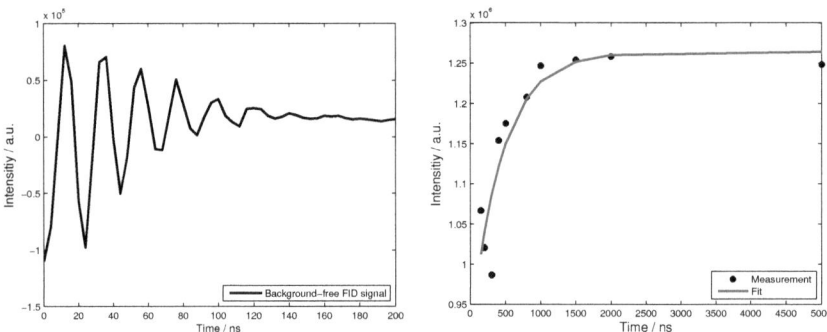

Figure A.3.: Left: FID signal of TEMPOL after the background subtraction for a delay time of 1000 ns. Right: Plot of the area of the magnitude of the background-free FID versus the delay time.

A.6. DNP Analysis

Calculation of the Enhancement

The DNP enhancement was calculated by dividing the sum of the NMR signal upon simultaneous microwave irradiation with the sum of the NMR signal without microwave irradiation (reference signal):

$$E = \frac{\sum \text{NMR data points with DNP}}{\sum \text{NMR data points without DNP}} = \frac{\sum_{i=1}^{N} y_i^{\text{DNP}}}{\sum_{i=1}^{N} y_i^{\text{Ref}}} = \frac{I^{\text{DNP}}}{I^{\text{Ref}}} \quad . \tag{A.3}$$

A. Appendix - Methods

Here, y_i denotes the amplitude of a data point in the NMR spectrum. I^{DNP} and I^{Ref} are the sum of the NMR signal. The error of the calculated sum was determined by using the propagation of uncertainty and taking the standard deviation of the baseline ($\Delta y_{\mathrm{noise}}$) which results in an uncertainty of each sum:

$$\Delta I^{\mathrm{DNP}} = \frac{\Delta y_{\mathrm{noise}}^{\mathrm{DNP}}}{\sqrt{N}} \quad , \tag{A.4}$$

$$\Delta I^{\mathrm{Ref}} = \frac{\Delta y_{\mathrm{noise}}^{\mathrm{Ref}}}{\sqrt{N}} \quad . \tag{A.5}$$

Subsequently, the error of the enhancement is

$$\Delta E = \pm \sqrt{\left(\frac{\Delta I^{\mathrm{DNP}}}{I^{\mathrm{Ref}}}\right)^2 + \left(\frac{I^{\mathrm{DNP}} \cdot \Delta I^{\mathrm{Ref}}}{(I^{\mathrm{Ref}})^2}\right)^2} \quad . \tag{A.6}$$

Extrapolated Enhancement

The maximum achievable enhancement E_{max} was computed by plotting the reciprocal Overhauser enhancement $(1-E)^{-1}$ against the reciprocal power P_{mw}^{-1} which should reveal a linear dependence. The data points in Figure A.4 demonstrate the expected linear relation. The intercept at $P^{-1} = 0$ of the linear regression yields the maximum enhancement when the EPR transition is saturated completely.

A.7. Determination of the Quality Factor of the EPR Probeheads

The quality factor Q was determined with the set-up shown in Figure A.5. The generated microwaves are lead to the resonator by the circulator (1 ⟶ 2). The microwaves reflected from the EPR cavity go to the detector which is a voltmeter (2 ⟶ 3). The measurement results in a plot of the voltage versus the microwave frequency which is fitted with a pseudo-Voigt function. The FWHM of the spectrum divided by the position of the extremum yields the quality factor Q. For example, the measurement of Q of the

A. Appendix - Methods

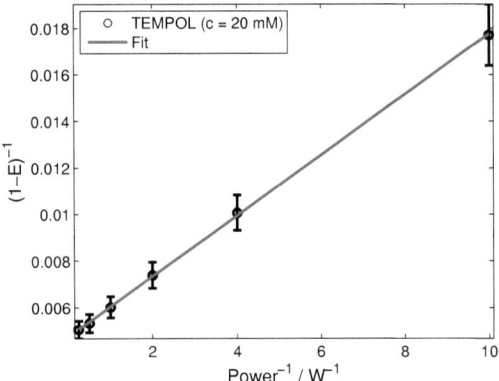

Figure A.4.: Plot of the reciprocal enhancement versus the reciprocal microwave power P_{mw} for TEMPOL ($c = 20$ mM). This plot demonstrates the expected linear dependence.

ENDOR probehead for a 3 mm tube, a capillary and without a sample are depicted in Figure A.6. The quality factors of the used probeheads are summarized in Table 3.3.

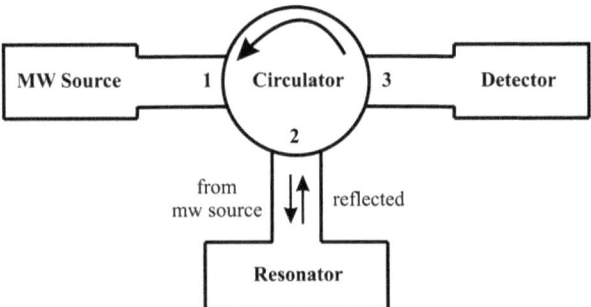

Figure A.5.: Circuit for the measurement of the quality factor. The circulator ensures that the microwave, which enters at point 1, can only exit at point 2 and the microwave entering at point 2 can only exit at point 3. In this case, the detector is a voltmeter which measures the reflection of the EPR probehead.

A. Appendix - Methods

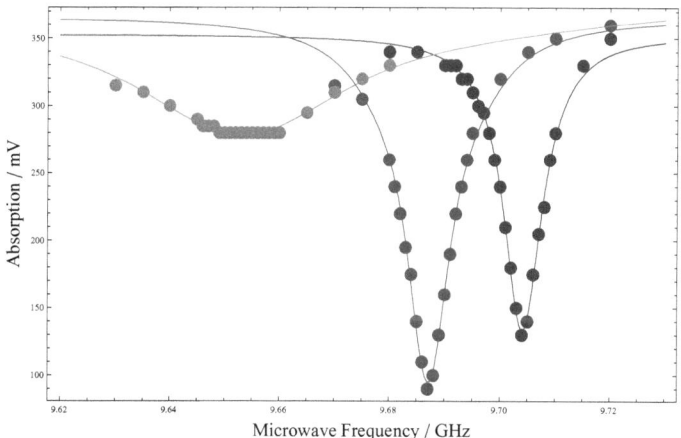

Figure A.6.: Q of the **ENDOR** probehead for a 3 mm tube (brown), a capillary (magenta) and without a sample (blue). The higher the loading of the probehead is the lower is the position of the minimum and the quality factor (*cf.* Table 3.3).

B. Materials - Polarizing Agents, Solvents and Used Molecules

B.1. TEMPO Derivatives

Materials

All TEMPO derivatives were used as received:

(i) 2,2,6,6-Tetramethylpiperidine-1-oxyl (TEMPO, Sigma).

(ii) 4-Hydroxy-2,2,6,6-tetramethylpiperidine-1-oxyl (TEMPOL, Sigma).

(iii) 4-Hydroxy-2,2,6,6-tetramethylpiperidine-1-^{15}N-oxyl (TEMPOL-^{15}N, Sigma).

(iv) 4-Hydroxy-2,2,6,6-tetramethylpiperidine-d_{17}-1-oxyl (TEMPOL-d_{17}, Sigma).

(v) 4-Hydroxy-2,2,6,6-tetramethylpiperidine-d_{17}-1-^{15}N-oxyl (TEMPOL-^{15}N-d_{17}, Sigma).

DNP Sample Preparation

For room temperature measurements, the TEMPO derivatives were dissolved in purified water (MilliQ system, Millipore) with a resistivity of 18.2 MΩcm. Subsequently, the sample was filled into capillaries (Blaubrand) with an inner diameter of 0.9 mm. The filling height ranged between 3 and 10 mm. The filling height strongly influences the achievable NMR signal enhancements as the B_{1e} in the ENDOR probehead is only homogeneous and non-zero in a volume of roughly $4 \times 4 \times 4$ mm^3. For low-temperature DNP measurements, the TEMPO derivatives were dissolved in glycerol-water mixture (60:40 vol%) as glycerol forms a glassy solution at low temperatures which prevents the radicals to migrate to the surface. Subsequently, the samples were shock-frozen in isobutane bathed in liquid nitrogen.

B.2. Spin-Labeled Heparin

Materials

Heparin (H4784) was purchased from Sigma. Trifluoroacetic anhydride, ethyl chloroformate, 1-Hydroxy-2,5-pyrrolidindion (NHS), 1-Ethyl-3-(3-dimethylaminopropyl)carbodiimid (EDC), TEMPO and 4-amino-TEMPO were purchased from Aldrich. All solvents used in this study were purified by standard procedures and distilled. 4-(carboxymethyl)-TEMPO was synthesized as described previously.[99] IR spectra were recorded on a Specord 75-IR spectrometer and UV spectra were recorded on a Specord M-40 spectrometer. X-band EPR spectra of the solutions of SL-heparins were recorded at room temperature on SE/X 2544 instrument using a microwave power of 2 mW, a modulation frequency of 100 kHz and an amplitude modulation of 0.032 mT. Mass spectroscopy studies were performed on a Finnigan-4021 (EI, 55 eV) apparatus. The progress of the reactions was monitored by high-performance liquid chromatography (HPLC), thin layer chromatography (TLC) and EPR techniques. HPLC analysis was performed on a Milikhrom chromatograph (column 2×64 mm, Separon C18 (5 mm), detection at 240 nm) using 30% aqueous MeCN containing KH_2PO_4 (0.05 M) as eluent. The retention volumes of 4-amino-TEMPO, amine **3** and ester **4** were 170, 185 and 1220 µl. TLC was carried out on Aldrich UV254 0.2-mm silica gel plates. Visualization of the TLC plates were performed by one of the following methods: (i) UV light (254 nm) or (ii) staining solution (0.3% ninhydrin in water containing 3% acetic acid) followed by heating.

Synthesis of Amine 3 and Ester 4

Synthesis of 4-[(5-aminopentyl)carbonylamino]-TEMPO, amine 3. A mixture of 1.31 g (10 mM) of 6-aminocaproic acid and 4.5 ml of trifluoroacetic anhydride was heated for 2 h at $80°C$ in soldered ampoule. Volatile components of the reaction mixture were evaporated at reduced pressure. The residual liquid consisted mainly of bis-trifluoroacetylated 6-aminocaproic acid. To achieve hydrolysis of mixed anhydride function of this intermediate, 0.25 ml (14 mM) of water was added to it with ice cooling. The solution was left for 1 h at $\sim 20°C$ and after that azeotroped with three 10 ml portions of dry benzene. The yield of 6-(trifluoroacetylamino)caproic acid was 2.27 g, mp $85-87°C$ (mp $88-90°C$[100]). It was dissolved in 10 ml of ethyl acetate and triethylamine (1.39 ml, 10 mM) and ethyl chloroformate (0.96 ml, 10 mM) were added sequentially at ice bath cooling and stirring. After stirring for 20 min at the same cooling, solution of 4-amino-TEMPO (1.71 g, 10 mM) in 2.5 ml of ethyl acetate was added in a 5 min time interval. The mixture was allowed to warm to room temperature and stirred for additional 30 min. The triethylammonium chloride salt was filtered off and washed with ethyl acetate (3 × 3 ml). The red ethyl acetate solution was washed sequentially with 2 ml of 0.1 M HCl, 1 ml of water and 2 ml of saturated NaHCO$_3$ solution and dried over anhydrous MgSO$_4$. Ethyl acetate was removed in vacuo to yield 3.9 g of 4-[(5-trifluoroacetylaminopentyl)carbonylamino]-TEMPO as a red oil. It was dissolved in 7 ml of ethanol and 11 ml of a 1 M NaOH solution in water was added at ∼ 10°C. The mixture was left at room temperature for 20 h. Solid K$_2$CO$_3$ (10 g) and 10 ml of ethyl acetate were added with stirring. The upper organic layer was separated, washed with saturated aqueous solution of NaCl, dried over anhydrous MgSO$_4$ and the solvent was removed in vacuo. The product was purified by column chromatography (silica gel, 5:1 to 2:1 chloroform/methanol) yielding 2.3 g of radical **3** as red hygroscopic solid which did not have a sharp melting point.

Synthesis of 4-(succinimidooxycarbonylmethyl)-TEMPO, ester 4. The product was obtained by modification of the known method.[101] To a mixture of 4-(carboxymethyl)-TEMPO (642 mg, 3 mmol) and NHS (345 mg, 3 mmol) in 6 ml of CH$_2$Cl$_2$ in dry atmosphere (Ar) under stirring at ice cooling was added DCC (618 mg, 3 mmol) in 3 ml of CH$_2$Cl$_2$ in a 15 min time interval. The solution was stirred for 1 h at ∼ 0°C and 12 h at ∼ 20°C. Precipitated DCU was filtered off and the solution was evaporated to dryness on a rotary evaporator. By re-crystallization of the residue from ethanol, pure ester **4** was obtained in a 75% yield (730 mg, 2.35 mmol). Orange crystals with m.p. 173 − 174.5°C (mp 157 − 159°C[101]).

Spin-Labeling of Heparin

Modification of heparin macromolecules by nitroxides was performed by means of amide bond formation.[102] For the determination of the fraction of modified disaccharides x (%), there exist three different approaches:

(i) *Approach based on recovery of nitroxide reagent.* The samples of reaction mixtures were taken and the unbound nitroxides were determined by HPLC and/or EPR after precipitation of the polymer by excess of MeCN. For heparin carboxyl group derivatization, the fraction x was calculated as $x = 100 \cdot 615 m/a$ (%), where 615 is the assumed average molecular weight of a disaccharide unit in unmodified heparin,

m is the quantity (mM) of 4-amino-TEMPO or 4-[H$_2$N(CH$_2$)$_5$C(O)NH]-TEMPO (3) bonded to a polymer and a is the quantity of heparin taken for the reaction (mg). For heparin amino groups' derivatization (after N-desulfation), the fraction of modified disaccharides was calculated (in assumption that 100% of amino groups were acylated) as $x = 100 \cdot 615m/(a + 102m)$ (%), where 102 is the difference in the molecular weight of the unmodified and N-desulfated disaccharide of heparin (-SO$_3$Na + H), m is the quantity (mM) of ester 4 bonded to a polymer and a is the quantity of N-desulfated heparin taken for the reaction (mg).

(ii) *Approach based on recovery of nitroxide reagent.* In UV spectra of SL-heparin derivatives, there is a band with maximum at ~ 245 nm which is due to nitroxide moiety absorption and it is practically coincident with the analogous band ($\lambda_{max} = 245$ nm) in the spectrum of TEMPO (spectrum not shown). The absorption of an equal concentration of unmodified heparin is negligibly small in this spectral range. Therefore, the fraction x of modified disaccharides was calculated as $x = (100DVM)/[\epsilon a l - DV(M' - M)]$, where M and M' are assumed as the average molecular weight of an unmodified disaccharide and the molecular weight of the modified disaccharide, D is the absorbance of the sample solution (a, mg) in V (ml) of water and $\epsilon = 1970$ l·mol·cm^{-1} is the molar extinction coefficient of TEMPO at 245 nm,[103] l is the cell thickness. Both methods (i) and (ii) coincide within error ≤ 10 relative %.

(iii) *Approach based on EPR spectra.* The degree of heparin derivation was determined by double integration of the EPR spectra of the modified heparin ($2 - 3$ mg/ml in water solutions) in comparison to the known concentration (0.5 mM in water) of the reference radical TEMPO (estimated accuracy ± 10%).

Coupling of 4-Amino-TEMPO with Heparin Carboxyl Groups (SL-heparin 1 and 4).

4-Amino-TEMPO (171 mg, 1 mmol) and NHS (115 mg, 1 mmol) were mixed with a solution of heparin sodium salt (615 mg, 1 mmol) in 10 ml of 0.1 M HCl. The solution obtained was cooled in an ice bath and EDC (230 mg, 1.2 mM) was added with stirring. The resulting solution was stirred in the ice bath for 30 min and after that stirring was continued at $\sim 20°C$. 4-Amino-TEMPO consumption was monitored by HPLC. Optimal pH $\sim 5 - 6$ was maintained by the adding of a 0.1 M NaOH solution. After 7 h 123 mg (0.72 mM) of 4-amino-TEMPO was coupled to heparin. The reaction mixture was freeze dried until its weight was ~ 4 g and absolute ethanol (30 ml) was added slowly with stirring. The precipitate was triturated to powder, filtered, washed with absolute ethanol (3 ml × 3), dissolved in water (3 ml) and re-precipitated with absolute ethanol (30 ml). The precipitate was washed with absolute ethanol and vacuum dried, yielding 660 mg of Hep-[C(O)NH-TEMPO]$_n$ (SLH-1) as a pale pink powder. The fraction of modified disaccharides x was found to be 72% (SL-heparin 4) by utilizing method (i). By UV determination (method (ii)), 67% of disaccharides bear TEMPO substitute.

Coupling of Amine 3 with Heparin Carboxyl Groups (SL-heparin 2).

The procedure was analogous to the previous one with a time of reaction of 24 h. The nitroxide radical content in the product was found to be 0.63 mmol/g ($x = 46\%$, method (i)). From the EPR spectrum of the derivative (method (iii)), x was found to be 42%.

Coupling of Ester 4 with Heparin Amino Groups (SL-heparin 3).

N-desulfation of heparin was performed by modification of described methods.[104-106] Sodium heparin (280 mg) was converted to the pyridinium salt by dissolving in water and passing through cationic exchange resin (Dowex 50WX4-400, H+ form) at the ice bath cooling. Double excess of pyridine was added relative to the total number of the acidic groups in heparin (final pH ~ 5). The solution obtained was frozen and lyophilized. The pyridinium heparin obtained (288 mg) was dissolved in 10 ml DMSO/water (95:5) and stirred in a 35°C water bath for 70 min. Ice-cooled water (10 ml) was added and the pH was adjusted to 9 with 0.2 N NaOH at ice-cooling. The solution was dialyzed against water (MWCO = 2000) and rotary evaporated at a 20°C water bath to a residual weight of ~ 0.7 g. Absolute ethanol (5 ml) was slowly added with mixing. The precipitate was filtered, washed with absolute ethanol (3 ml × 2) and vacuum dried, yielding 203 mg of partially N-desulfated heparin Hep-(NH$_2$)$_n$. It was dissolved in 10 ml of DMSO/H$_2$O (2:1), saturated with NaHCO$_3$. 124 mg (0.40 mM) of 4-(succinimidooxycarbonylmethyl)-TEMPO was pounded with a pestle and added to the Hep-(NH$_2$)$_n$ solution with stirring (~ 1 equiv of ester 4 per 1 equiv of disaccharide unit). Mixing was continued at $\sim 20°C$ and pH 8−8.5 until no more consumption of ester 4 took place. By HPLC of the reaction mixture, it was found that 0.23 (12 h), 0.24 (24 h) and 0.24 (36 h) mmol of ester 4 was coupled to heparin and ~ 0.14 mmol of ester 4 still existed in the mixture after 36 h reaction time. From these data it was concluded that all free amino groups of Hep-

B. Materials - Polarizing Agents, Solvents and Used Molecules

$(NH_2)_n$ were acylated. The volume of reaction mixture was halved by freeze drying and the polymer was precipitated by addition of excess of acetone/ether (1:1), centrifuged, decanted and washed with acetone. Dried polymer was dissolved in water (1.5 ml), the solution was centrifuged, decanted and re-precipitated with absolute ethanol (15 ml). The precipitate was washed with absolute ethanol and vacuum dried, yielding 189 mg of Hep-[NHC(O)CH$_2$-TEMPO]$_n$ (2) as a pale orange powder. The nitroxide radical content in the product was found to be 0.96 mmol/g ($x = 65\%$, method (i)). By UV determination (method (ii)) 61% of disaccharides bear TEMPO substitute.

Sample Preparation

For room and low-temperature measurements, the radicals are dissolved in water or water/glycerol solutions as described before. The exact mass of the SL-heparins and the equivalent radical concentrations when dissolved in solution are given in Table B.1.

Table B.1.: Molecular weight of the SL-heparins and the equivalent radical concentration when dissolved in solution.

Radical	SL-heparin 1	SL-heparin 2	SL-heparin 3	SL-heparin 4
labeling degree x [%]	18	45	65	72
M$_W$ [g/mol]	627.7	725.7	666.5	697.5
concentration [mM]	20	20	20	20
Volume [µl]	200	200	200	200
m [mg]	31.4	6.5	3.9	4.1

B.3. SL-Hydrogel

Materials

Methacrylic acid (MAA, 99%, Aldrich) was distilled prior to use. N,N-Ethylmethylacrylamide (EMAAm) and 4-Methacryloylbenzophenone (MABP) were prepared according to reported procedures.[107,108] Dioxane was distilled over CaH_2 and dried over a molecular sieve (4 Å). 2,2'-Azobis(isobutyronitrile) (AIBN, 98%, Acros) was recrystallized from methanol. Triethylamine (NEt3, 99.5%, Fluka), pentafluorophenyltrifluoroacetate (PFTFA, 98%, Aldrich) and 4-amino-2,2,6,6-tetramethylpiperidine-1-oxyl (Amino-

TEMPO, 97%, Acros) were used as received. Distilled water was further purified by a MilliQ System (Millipore) to achieve a resistivity of 18.2 MΩcm.

Thermoresponsive Polymer

Statistical terpolymers based on EMAAm (3 g, 26.5 mmol) admixed with small amounts of MAA and MABP were obtained via free radical polymerization initiated by AIBN (20 mg, 0.12 mmol). The reaction took place in 20 mL dioxane at 60 °C for 24 h under exclusion of air and moisture. The polymers were precipitated in 200 mL diethyl ether and purified by re-precipitation from ethanol into diethyl ether. They were freeze-dried from water in vacuo. The yield was around 90%. ^{1}H-NMR (250 MHz, d^4-$MeOH$): δ / ppm = 0.9-1.2 (m, $-CH_3$, MAA, MABP, $-CH_2CH_3$ EMAAm), 1.5-1.9 (m, CH_2 backbone), 2.2-3.7 (m, CH backbone, $-CH_2CH_3$, $-Me$, EMAAm), 6.8-8.0 (m, C-H_{arom}). The molecular weight and the polydispersity index (M_w/M_n) were determined by gel permeation chromatography with dimethylformamide as mobile phase. The measurement was conducted at 60°C, PMMA served as internal standard. The monomer composition and the molecular weight distribution of the synthesized polymers are summarized in Table B.2.

Table B.2.: Properties of the Synthesized Polymers.

	monomer composition [mol%]					
	EMAAm	MAA	MABP	M_w / g mol$^{-1 a}$	$M_w/M_n{}^a$	Yield / %
P5	94	5	1	49,600	3.1	89
P15	84	15	1	28,000	4.2	92

^{a} determined by GPC at 60°C in DMF

Spin-Labeling and Sample Preparation

15 wt% polymer solutions in ethanol were drop-cast on hexamethyldisilazane modified glass slides and dried over night in vacuo at 40°C to achieve polymeric films in the range of $10 - 15\,\mu$m. The polymer was crosslinked and tethered to the substrate by UV irradiation (6.82 J/cm^2) with a Stratagene UV Stratalinker with a peak wavelength of 365 nm. The carboxylic acid groups of the resulting gels were converted to active ester units with a two-fold excess of PFTFA and NEt_3 in dichloromethane (DMC). After

washing the gels with DMC twice, they were reacted with a two-fold excess of 4-amino-TEMPO and NEt_3 in DMC to achieve the spin-labeled material. The synthetic route is described in Figure B.1.

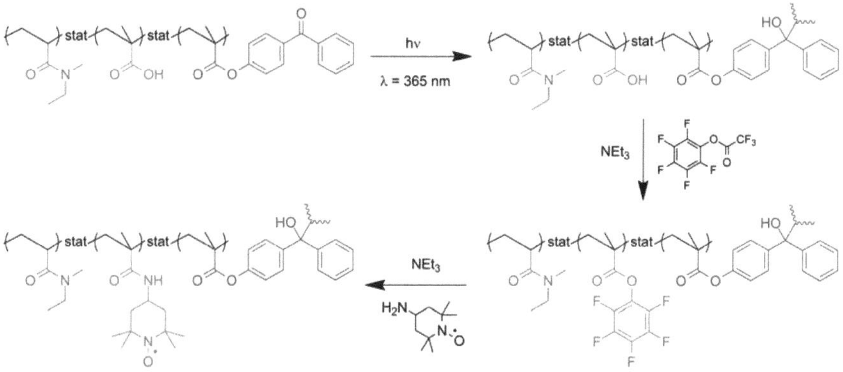

Figure B.1.: Synthetic route to the spin-labeled hydrogel.

In order to remove unreacted nitroxide molecules, the gels were subjected to dialysis (Spectra/Por 3 Dialysis Membrane, MWCO = 3500 gmol^{-1}) in ethanol for one week. The solvent was removed in vacuo and the spin-labeled gel was abraded with a scalpel, filled into EPR sample canules (i.d. 0.8 mm) and swollen with deionized water. Excess water not bound to the hydrogel was removed resulting in a filling height of approximately 5 − 8 mm. For low-temperature measurements, we used 3 mm o.d. sample tubes. Subsequently, the swollen SL-hydrogels were shock-frozen in isobutane in a liquid nitrogen bath to prevent a phase separation.

B.4. Solvents

Hexafluorobenzene

Hexafluorobenzene (HFB) was purchased from Aldrich and used as received. It is liquid at room temperature and has a density of 1.6 g/ml. Noteworthy, the dielectric constant at room temperature is much lower as compared to water so that large sample volumes of hexafluorbenzene can be inserted into the EPR probehead without a severe influence on the EPR mode. The dielectric constant of hexafluorobenzene is not known but the dielectric constant of benzene is approximately $\epsilon_r = 2-3$ as compared to water $\epsilon_r = 80$ at 20 °C. This property makes it an ideal solvent for the investigation of DNP from the EPR point of view.

B.5. Solutes

KJT (^{19}F)

KJT is a statistical, hyperbranched polymer of dimethyaminoethylacrylate (DMAEA, 77 mol%) and trifluoroethylacrylate (tFEA, 19 mol%). It was synthesized using ethyleneglycol dimehtacrylate (EGDMA, 4 mol%) as branching agent. The molecular weight as determined by GPC is 22 kDa with a polydispersity of 1.8. On average, the hyperbranched polymers were 8-arm stars. The T_{2n} time of ^{19}F at 7 T was 68 ms. the hydrodynamic radius in water at 25 °C was 8.6 nm as determined by dynamic light scattering. All calculations were done at a concentration of 20 mg/ml in water. Data shown were provided by Dr Kristofer Thurecht (Centre for Advanced Imaging, University of Queensland, St Lucia, Queensland, Australia).

Urea (^{13}C)

^{13}C-labeled urea was purchased from Aldrich and used as received. It is solid at room temperature and has a molecular weight of ~ 60 g/mol. The solubility in water is very good with approximately 1000 g/l at 20 °C. For DNP investigations 180 mg urea were dissolved in 300 μl de-ionized water ($R \geq 18.2$ Ωcm) resulting in a concentration of 10 mol/l.

C. Matlab Scripts

In this Chapter, the home-written Matlab programs are displayed.

C.1. Enhancement and Power Dependence

Here, the calculation of the enhancement and the simultaneous determination of the extrapolated enhancement are given.

```
1  clear all; close all;
2
3  pathnameA = 'F:\DNP_Daten\200911_November\091120_TEMPOL_20mM_288Kb\
       TEMPOLref\';
4  pathnameB = 'F:\DNP_Daten\200911_November\091120_TEMPOL_20mM_288Kb\
       TEMPOLpow\';
5
6  power = [.1 0.25 0.5 1 2 4 8 16]; %irradiated power in Watt EPRfreq=9
       .7025GHz
7  off1 = 0;
8  off2 = 0;
9  pts = 6; % looks in interval +-pts for maximum
10 ptswidth = 5; % autophase in interval +-ptswidth
11 intli = 3; % integration from intli
12 intre = 4; % to intre
13 %% mean area of reference
14 npts = 12;
15 A = zeros(npts,4);
16 for n=1:npts
17
18    % Load the acquisition parameters first
19    nn = num2str(n);
20    fname = [pathnameA nn,'\acqu.par'];
21    para = readkea2_par(fname);
22    dwell = para.acq.dw;              % dwell time in s
23    npts1 = para.acq.pts;             % number of data points
24    freqax = para.freqax;             % Frequenzachse in Hz
```

```matlab
scans = para.acq.scans                  % No of reference scans

% Load data with readkea

a = [pathnameA nn,'\data.csv'];
b = num2str(a);
c = readkea_var_bcd(off1,pts,ptswidth,b);

% Calculate integral of the peak

[ma,ind] = max(real(c.fftdata(npts1/2-pts+off1:npts1/2+pts+off1,1)));
ind = ind + (npts1/2-(pts+1)+off1);
  %ma=1;

intmin = int16(ind)-intli;
intmax = int16(ind)+intre;

V = real(c.fftdata(intmin:intmax,1));
integ = trapz(V)/scans;

%   figure;
%   plot(freqax(intmin:intmax),V);

A(n,1) = n;
A(n,2) = ma;
A(n,3) = ind;
A(n,4) = integ;

end

NN = find(A(:,4)~=0);
mittel = mean(A(NN,4))%
fehler = std(A(NN,4))

npoints = 6;
B = zeros(npoints,5);
E = zeros(npoints,3);
for n=1:npoints
%    if n>5
%       m=n+1;
%    else
    m=n;
```

```matlab
67  %   end
68
69      % Load the acquisition parameters first
70      nn = num2str(m);
71      fname2 = [pathnameB nn,'\acqu.par'];
72      para2 = readkea2_par(fname2);
73      dwell2 = para2.acq.dw;              % dwell time in s
74      npts2 = para2.acq.pts;              % number of data points
75      freqax2 = para2.freqax;             % Frequenzachse in Hz
76      NMRfreq2 = para2.acq.freq;          % NMR observe frequency
77      scans2 = para2.acq.scans            % No of DNP scans
78
79      % Load data with readkea
80
81      a2 = [pathnameB nn,'\data.csv'];
82      b2 = num2str(a2);
83      c2 = readkea_var_bcd(off2,pts,ptswidth,b2);
84
85  % Calculate integral of the peak
86
87      [ma2,ind2] = max(real(c2.fftdata(npts2/2-pts+off2:npts2/2+pts+off2,1)
            ));
88      ind2 = ind2 + (npts2/2-(pts+1)+off2);
89        %ma=1;
90
91      intmin2 = int16(ind2)-intli;
92      intmax2 = int16(ind2)+intre;
93
94      V2 = real(c2.fftdata(intmin2:intmax2,1));
95      integ2 = trapz(V2)/scans2;
96
97  %   figure;
98  %     plot(freqax2(intmin2:intmax2),V2);
99
100     B(n,1) = n;
101     B(n,2) = ma2;
102     B(n,3) = ind2;
103     B(n,4) = integ2;
104     B(n,5) = NMRfreq2;
105     E(n,2) = integ2/mittel; % Calculated enhancement 1 to 1
106     enhance = E(n,2)
107     delta_enhance = fehler*integ2/mittel^2
```

```matlab
108    E(n,3) = fehler*integ2/mittel^2;
109 end
110
111 MM = find(E(:,2)~=0);
112 E(MM,1) = power(MM);
113 % evaluation
114 E = sortrows(E,1);
115
116 [data] = [[.25 0.5 1 2]' E(2:5,2)]
117 fleakage = .86;
118 rho = 0.05;
119 alpha = 10;
120 % Put parameters to be floated into vector P
121 P = [alpha rho];   %
122
123 [myopt]=optimset('Display','final','MaxIter',1e7,'MaxFunEvals',1e7,'
        TolFun',1e-7,'TolX',1e-7);% 'iter'
124
125     [Popt, f_out, exitflag] = fminsearch(@minimizer_Emax, P, myopt,
            data, fleakage)
126     testf = minimizer_Emax_values(Popt, E(:,1), fleakage)
127
128 fprintf(['alpha = ' num2str(Popt(1)) '\n']);
129 fprintf(['E_{max} = ' num2str(1-658.21.*Popt(1,2).*fleakage) '\n\n']);
130
131 figure(1004);
132 errorbar(E(MM,1), -E(MM,2), E(MM,3), 'ok', 'LineWidth',2); %'Color', 'k
        ', 'Marker','o','MarkerSize',6
133 hold on
134 plot(E(MM,1), -testf, '-r', 'LineWidth',2);
135 %title('Enhancement vs Power - Saturation Factor','FontSize',16);
136 legend('TEMPOL (c = 20 mM)', 'Fit', 'Location', 'NorthEast');
137 set(gcf,'Color',[1,1,1]);
138 set(gca,'FontSize',12);
139 xlabel('Power / W','FontSize',14);
140 ylabel('Negative enhancement','FontSize',14);
141 % axis([0.0 1.1 0.01 0.1]);
142 ylim = get(gca, 'YLim');
143 text(-1.5, ylim(2), '(a)','FontSize',18);
144
145 %export_fig 'F:\PhD\LaTex\PhD_Arbeit\pics\lindep_TEMPOL_20mMa.pdf' -
        r600 -pdf
```

```matlab
% Calculation of Emax
% Enhancement factors
enhance = E(MM,2)*-1;
% 1/P
rec_power = 1./E(MM,1);
% 1/1-E
sat_enhance=1./(1-enhance);
delta_sat_enhance = E(:,3).*sat_enhance.^2;

% now fit the enhancement data

% linear decay, sample1
    [p_sample,s_sample]=polyfit(rec_power,sat_enhance,1);
    sat_enhance_fit=polyval(p_sample,rec_power); %creates baseline from
         10% of points in lf and 10% of points in hf data range
    E_max=1-1/p_sample(2);
        [m,b,dm,db,chi2] = linfit(rec_power,sat_enhance,
            delta_sat_enhance);% for y=m*x+b and chi squared
    Emax_linfit = 1-1/b
    delta_Emax2 = db/b^2;
    fprintf(['\n fit values for sample saturation: 1/1-E_{max} = ',
        num2str(p_sample(2)),' ',num2str(p_sample(1)) 'W * 1/P \n']);
    fprintf([' E_{max}_sample=',num2str(E_max),' +- ',num2str(
        delta_Emax2),'\n\n']);

figure(1003);
errorbar(rec_power, sat_enhance, delta_sat_enhance, 'ok', 'LineWidth',
    2);
hold on
plot(rec_power, sat_enhance_fit, '-r', 'LineWidth', 2)
%title('Enhancement vs Power - Saturation Factor','FontSize',16);
legend('TEMPOL (c = 20 mM)', 'Fit', 'Location', 'NorthWest');
% sample1=['E_{max}TEMPOL = ' num2str(round(E_max))];
% text(1,.01,sample1,'FontSize',10);
% sample2=['E(16W) Hydrogel 15% = ' num2str(round(E(1,2)))];
% text(1,.07,sample2,'FontSize',10);
set(gcf,'Color',[1,1,1]);
set(gca,'FontSize',12);
xlabel('Power^{-1} / W^{-1}','FontSize',14);
ylabel('(1-E)^{-1}','FontSize',14);
axis tight
```

```
183  text(-1.8, 0.022, '(b)','FontSize',18);
184
185  %export_fig 'F:\PhD\LaTex\PhD_Arbeit\pics\lindep_TEMPOL_20mM.pdf' -r600
        -pdf
```

The output of the program is presented in Figure C.1.

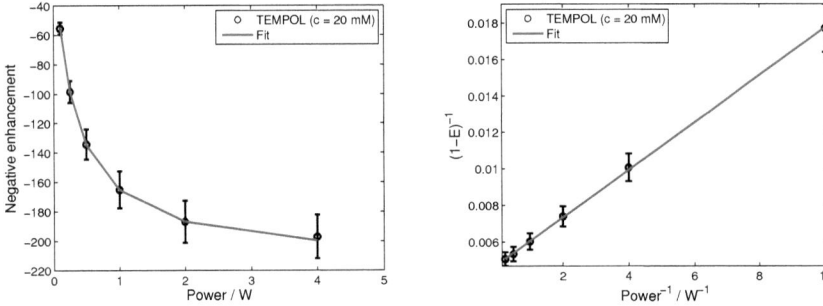

Figure C.1.: Output of the program given above. On the left, the enhancement versus the power is plotted. On the right hand side, the inverse enhancement is plotted against the inverse microwave power which results in a linear dependence.

C.2. Magnetic Field Sweep

Here, the two programs for the evaluation of the magnetic field sweep at room temperature is shown for several microwave powers and the appropriate CW EPR spectrum of TEMPOL ($c = 0.5$ mM). The resulting Figure C.2 can be found below.

```matlab
clear all; close all;
%pathnameA = 'C:\Dokumente und Einstellungen\DNP-Group\Eigene Dateien\
    Dollmann\200808 August\080821_PROXYL_03mM\PROXYLref\';
pathnameA = 'F:\DNP_Daten\201004_April\100429_TEMPOL_0p5mM_Bsweep\
    TEMPOLref\';
pathnameB = 'F:\DNP_Daten\201004_April\100429_TEMPOL_0p5mM_Bsweep\
    TEMPOLpow';

EPRfreq = 9.76; %Microwave frequency in GHz

off = 0;
pts = 10; % looks in interval +-pts for maximum
ptswidth = 4; % autophase in interval +-ptswidth
intli = 4; % integration from intli
intre = 4; % to intre
%% calculate enhancement
numpow = 4;
npts=47;%47
A = zeros(npts,4,numpow);
B = zeros(npts,5,numpow);
E = zeros(npts,3,numpow);
for lei=1:numpow
for n=1:npts
    % Load the acquisition parameters first

    nn = num2str(n);
    fname1 = [pathnameA nn,'\acqu.par'];
    fname2 = [pathnameB num2str(lei) '\' nn,'\acqu.par'];
    para1 = readkea2_par_091210(fname1);
    para2 = readkea2_par_091210(fname2);
    dwell1 = para1.acq.dw;                  % dwell time in s
    dwell2 = para2.acq.dw;                  % dwell time in s
    npts1 = para1.acq.pts;                  % number of data points
    npts2 = para2.acq.pts;                  % number of data points
```

C. Matlab Scripts

```matlab
32    freqax1 = para1.freqax;               % Frequenzachse in Hz
33    freqax2 = para2.freqax;               % Frequenzachse in Hz
34    scans1 = para1.acq.scans;             % No of reference scans
35    scans2 = para2.acq.scans;             % No of DNP scans
36    NMRfreq2 = para2.acq.freq;            % NMR observe frequency
37
38    % Load data with readkea
39
40    a2 = [pathnameB num2str(lei) '\' nn,'\data.csv'];
41    b2 = num2str(a2);
42    c2 = readkea_var_bcd(off,pts,ptswidth,b2);
43    a1 = [pathnameA nn,'\data.csv'];
44    b1 = num2str(a1);
45    c1 = readkea_var_bcd(off,pts,ptswidth,b1);
46
47    % Calculate integral of the peak
48
49    [ma1,ind1] = max(real(c1.fftdata(npts1/2-pts+off:npts1/2+pts+off,1)))
           ;
50    ind1 = ind1 + (npts1/2-(pts+1)+off);
51    [ma2,ind2] = max(real(c2.fftdata(npts2/2-pts+off:npts2/2+pts+off,1)))
           ;
52    ind2 = ind2 + (npts2/2-(pts+1)+off);
53      %ma=1;
54
55    intmin1 = int16(ind1)-intli;
56    intmax1 = int16(ind1)+intre;
57    intmin2 = int16(ind2)-intli;
58    intmax2 = int16(ind2)+intre;
59
60    V1 = real(c1.fftdata(intmin1:intmax1,1))/scans1;%;
61    integ1 = trapz(V1);
62    V2 = real(c2.fftdata(intmin2:intmax2,1))/scans2;%;
63    integ2 = trapz(V2);
64
65  %   figure;
66  %   plot(freqax1(intmin1:intmax1),V1);
67  %   figure;
68  %   plot(freqax2(intmin2:intmax2),V2);
69
70    A(n,1,lei) = n;
71    A(n,2,lei) = ma1;
```

```matlab
72    A(n,3,lei) = ind1;
73    A(n,4,lei) = integ1;
74    B(n,1,lei) = n;
75    B(n,2,lei) = ma2;
76    B(n,3,lei) = ind2;
77    B(n,4,lei) = integ2;
78    B(n,5,lei) = EPRfreq;
79    E(n,1,lei) = NMRfreq2;
80    Bfield2 = NMRfreq2/0.0042576375;
81    E(n,2,lei) = Bfield2;
82    diff = abs(c1.phase-c2.phase);
83 %    if diff<280 && diff>90
84 %        E(n,3) = -integ2/integ1; % Calculated enhancement 1 to 1
85 %    else
86      E(n,3,lei) = integ2/integ1;
87 %    end
88    enhance = E(n,3,lei)
89
90  end
91  E(:,:,lei) = sortrows(E(:,:,lei),1);
92 end
93
94 save (['F:\DNP_Daten\201004_April\100429_TEMPOL_0p5mM_Bsweep\' '
        E_all_sweep.mat'], 'E', '-mat');
```

```matlab
1  clear all; close all;
2
3  % load DNP sweep
4  pathnameA = 'F:\DNP_Daten\201004_April\100429_TEMPOL_0p5mM_Bsweep\';
5  E = load([pathnameA, 'E_all_sweep.mat'], '-mat'); E=E.E;
6
7  % load EPR spectrum
8  pathname='F:\DNP_Daten\EPR_Daten\200910_Oktober\091023_TEMPOL\';
9  [B, TEMPOL, par] = eprload([pathname '091023_TEMPOL_0p5mM.dta']);%disp(
       par)
10 points = par.XPTS; xmini = par.XMIN/10; xwidth = par.XWID/10; xcenter =
        xmini + xwidth/2;
11 bsldat = round(points/10); %creates baseline from 10% of points in lf
       and 10% of points in hf data range
12 bsli = [1:bsldat (points-bsldat):points];
```

```matlab
[p,s] = polyfit(B(bsli),TEMPOL(bsli),1);
base = polyval(p,B);%this is the baseline
TEMPOL = TEMPOL-base;% sloppy baseline correction;-base
TEMPOL = TEMPOL-mean(TEMPOL);% sloppy baseline correction;-base
ABS_TEMPOL = cumsum(TEMPOL);

B=B-B(find(ABS_TEMPOL==max(ABS_TEMPOL)));

center = find(E(:,3,1)==max(E(:,3,1)));
sweepax=(E(:,2,1)-E(center,2,1));

Bneu = B(find(B>min(sweepax) & B<max(sweepax)));
ABS_TEMPOLneu = interp1(B, ABS_TEMPOL, Bneu);
ABS_TEMPOLneu = ABS_TEMPOLneu/max(ABS_TEMPOLneu)*max(max(E(:,3,:)));
%% now plot
figure;
hold all
plot(Bneu, ABS_TEMPOLneu, 'k', 'LineWidth',2);
plot(sweepax, E(:,3,1), '--b', 'LineWidth',2);
plot(sweepax, E(:,3,2), '--g', 'LineWidth',2);
plot(sweepax, E(:,3,3), '--m', 'LineWidth',2);
plot(sweepax, E(:,3,4), '--r', 'LineWidth',2);
%axis ([3430 3550 -50 50]);
axis tight;
xlabel('magnetic field offset / Gauss');
ylabel('negative enhancement');
title('Magnetic field sweep TEMPOL 0.5mM');
legend('CW ABS', '0.5 W', '1 W', '4 W', '8 W')
grid on
box on
set(gcf,'Color',[1,1,1])
set(gca,'FontSize',12)

%export_fig 'F:\PhD\LaTex\PhD_Arbeit\pics\Bsweep.pdf' -r600 -pdf
```

C. *Matlab Scripts*

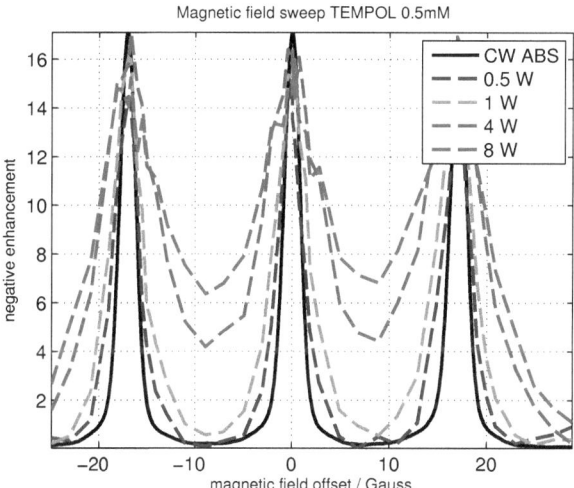

Figure C.2.: Visual output of the program given above. The magnetic field sweep at room temperature for several microwave powers and the appropriate CW EPR spectrum of **TEMPOL** ($c = 0.5$ mM) are plotted against the magnetic field offset.

Bibliography

[1] Bloch, F. *Phys. Rev.* **1946**, *70*, 460.

[2] Purcell, E. M.; Torrey, H. C.; Pound, R. V. *Phys. Rev.* **1946**, *69*, 37.

[3] Bowers, C. R.; Weitekamp, D. P. *Journal of the American Chemical Society* **1987**, *109*, 5541–5542.

[4] Natterer, J.; Bargon, J. *Progress in Nuclear Magnetic Resonance Spectroscopy* **1997**, *31*, 293–315.

[5] Adams, R. W.; Aguilar, J. A.; Atkinson, K. D.; Cowley, M. J.; Elliott, P. I. P.; Duckett, S. B.; Green, G. G. R.; Khazal, I. G.; Lopez-Serrano, J.; Williamson, D. C. *Science* **2009**, *323*, 1708–1711.

[6] Hore, J.; Broadhurst, R. W. *Progress in Nuclear Magnetic Resonance Spectroscopy* **1993**, *25*, 345–402.

[7] Bargon, J. *HCA* **2006**, *89*, 2082–2102.

[8] Daviso, E.; Diller, A.; Alia, A.; Matysik, J.; Jeschke, G. *Journal of Magnetic Resonance* **2008**, *190*, 43–51.

[9] Kastler, A. *Journal De Physique Et Le Radium* **1950**, *11*, 255–265.

[10] Bouchiat, M. A.; Carver, T. R.; Varnum, C. M. *Phys. Rev. Lett.* **1960**, *5*, 373–.

[11] Abragam, A. *The principles of nuclear magnetism.*; Clarendon Press: Oxford, 1961.

[12] Abragam, A.; Goldman, M. *Nuclear magnetism: Order and disorder;* Clarendon Press: Oxford, 1982.

[13] Ardenkjaer-Larsen, J. H.; Fridlund, B.; Gram, A.; Hansson, G.; Hansson, L.; Lerche, M. H.; Servin, R.; Thaning, M.; Golman, K. *Proceedings of the National Academy of Sciences* **2003**, *100*, 10158–10163.

[14] Münnemann, K.; Bauer, C.; Schmiedeskamp, J.; Spiess, H. W.; Schreiber, W. G.; Hinderberger, D. *Applied Magnetic Resonance* **2008**, *34*, 321–330.

[15] Dollmann, B. C.; Junk, M. J. N.; Drechsler, M.; Spiess, H. W.; Hinderberger, D.; Munnemann, K. *Phys. Chem. Chem. Phys.* **2010**, *12*, 5879–5882.

[16] Golman, K.; Zandt, R. i.; Lerche, M.; Pehrson, R.; Ardenkjaer-Larsen, J. H. *Cancer Research* **2006**, *66*, 10855–10860.

[17] Golman, K.; Petersson, J. S.; Magnusson, P.; Johansson, E.; Akeson, P.; Chai, C.-M.; Hansson, G.; Mansson, S. *Magnetic Resonance in Medicine* **2008**, *59*, 1005–1013.

[18] Atherton, N. M. *Ellis Horwood series in physical chemistry;* Ellis Horwood: PTR Prentice Hall: New York, 1993.

[19] Schweiger, A.; Jeschke, G. *Principles of pulse electron paramagnetic resonance;* Clarendon Press: Oxford, 2001.

[20] Slichter, C. P. *Principles of Magnetic Resonance;* Springer-Verlag: Berlin, 1980.

[21] Callaghan, P. T. *Principles of Nuclear Magnetic Resonance Microscopy;* Oxford University Press: Oxford, 1991.

[22] Hahn, E. L. *Phys. Rev.* **1950**, *80*, 580.

[23] Carr, H. Y.; Purcell, E. M. *Phys. Rev.* **1954**, *94*, 630.

[24] Pannier, M.; Veit, S.; Godt, A.; Jeschke, G.; Spiess, H. W. *Journal of Magnetic Resonance* **2000**, *142*, 331–340.

[25] Pake, G. E. *J. Chem. Phys.* **1948**, *16*, 327–336.

[26] Jeschke, G.; Polyhach, Y. *Phys. Chem. Chem. Phys.* **2007**, *9*, 1895–1910.

[27] Jeschke, G.; Chechik, V.; Ionita, P.; Godt, A.; Zimmermann, H.; Banham, J.; Timmel, C. R.; Hilger, D.; Jung, H. *Applied Magnetic Resonance* **2006**, *30*, 473–498.

[28] Banham, J.; Baker, C.; Ceola, S.; Day, I.; Grant, G.; Groenen, E.; Rodgers, C.; Jeschke, G.; Timmel, C. *Journal of Magnetic Resonance* **2008**, *191*, 202–218.

[29] Lawler, R. G. *Accounts of Chemical Research* **1972**, *5*, 25–33.

[30] Song, C.; Hu, K.-N.; Joo, C.-G.; Swager, T.; Griffin, R. *Journal of the American Chemical Society* **2006**, *128*, 11385-11390.

[31] Sezer, D.; Gafurov, M.; Prandolini, M. J.; Denysenkov, V. P.; Prisner, T. F. *Phys. Chem. Chem. Phys.* **2009**, *11*, 6638–6653.

[32] McCarney, E. R.; Armstrong, B. D.; Lingwood, M. D.; Han, S. *Proceedings of the National Academy of Sciences* **2007**, *104*, 1754–1759.

[33] Höfer, P.; Parigi, G.; Luchinat, C.; Carl, P.; Guthausen, G.; Reese, M.; Carlomagno, T.; Griesinger, C.; Bennati, M. *Journal of the American Chemical Society* **2008**, *130*, 3254–3255.

[34] Comment, A.; van den Brandt, B.; Uffmann, K.; Kurdzesau, F.; Jannin, S.; Konter, J.; Hautle, P.; Wenckebach, W.; Gruetter, R.; van der Klink, J. *Concepts in Magnetic Resonance Part B: Magnetic Resonance Engineering* **2007**, *31B*, 255–269.

[35] Overhauser, A. W. *Phys. Rev.* **1953**, *92*, 411.

[36] Abragam, A. *Phys. Rev.* **1955**, *98*, 1729.

[37] Carver, T. R.; Slichter, C. P. *Phys. Rev.* **1956**, *102*, 975–.

[38] Abragam, A.; Goldman, M. *Reports on Progress in Physics* **1978**, *41*, 395–467.

[39] Maly, T.; Debelouchina, G. T.; Bajaj, V. S.; Hu, K.-N.; Joo, C.-G.; MakJurkauskas, M. L.; Sirigiri, J. R.; van der Wel, P. C. A.; Herzfeld, J.; Temkin, R. J.; Griffin, R. G. *J. Chem. Phys.* **2008**, *128*, 052211–19.

[40] Hausser, K. H.; Stehlik, D. *Advanced Magnetic Resonance* **1968**, *3*, 79–139.

[41] Turke, M.-T.; Tkach, I.; Reese, M.; Hofer, P.; Bennati, M. *Phys. Chem. Chem. Phys.* **2010**, *12*, 5893–5901.

[42] Solomon, I. *Phys. Rev.* **1955**, *99*, 559.

[43] Freed, J. H. *Journal Of Chemical Physics* **1978**, *68*, 4034–4037.

[44] Hwang, L. P.; Freed, J. H. *Journal Of Chemical Physics* **1975**, *63*, 4017–4025.

[45] Ayant, Y.; Belorizky, E.; Aluzon, J.; Gallice, J. *J. Phys. France* **1975**, *36*, 991–1004.

[46] Brunner, H.; Hausser, K. H. *Journal of Magnetic Resonance (1969)* **1972**, *6*, 605–611.

[47] Lingwood, M. D.; Han, S. *Journal of Magnetic Resonance* **2009**, *201*, 137–145.

[48] Levitt, M. H. *Spin dynamics basics of nuclear magnetic resonance;* John Wiley & Sons: Chichester; New York, 2001.

[49] Duijvestijn, M. J.; Wind, R. A.; Smidt, J. *Physica B+C* **1986**, *138*, 147–170.

[50] Jeffries, C. D. *Phys. Rev.* **1957**, *106*, 164.

[51] Wenckebach, W. T. *Applied Magnetic Resonance* **2008**, *34*, 227–235.

[52] Hwang, C. F.; Hill, D. A. *Phys. Rev. Lett.* **1967**, *19*, 1011.

[53] Wollan, D. S. *Phys. Rev. B* **1976**, *13*, 3671.

[54] Halbach, K. *Nuclear Instruments and Methods* **1980**, *169*, 1–10.

[55] Stevenson, S.; Dorn, H. C. *Analytical Chemistry* **1994**, *66*, 2993–2999.

[56] Reese, M.; Lennartz, D.; Marquardsen, T.; Hofer, P.; Tavernier, A.; Carl, P.; Schippmann, T.; Bennati, M.; Carlomagno, T.; Engelke, F.; Griesinger, C. *Applied Magnetic Resonance* **2008**, *34*, 301–311.

[57] Danieli, E.; Mauler, J.; Perlo, J.; Blümich, B.; Casanova, F. *Journal of Magnetic Resonance* **2009**, *198*, 80–87.

[58] Jagschies, L. "Aufbau eines 6K Polarisators zur Hyperpolarisation von Kontrastmitteln für die Magnetresonanztomographie", Master's thesis, Max Planck Institute for Polymer Research, University of Mainz, 2009.

[59] Gerfen, G. J.; Becerra, L. R.; Hall, D. A.; Griffin, R. G.; Temkin, R. J.; Singel, D. J. *The Journal of Chemical Physics* **1995**, *102*, 9494-9497.

[60] Hu, K.-N.; Yu, H.-h.; Swager, T.; Griffin, R. *Journal of the American Chemical Society* **2004**, *126*, 10844-10845.

[61] Ardenkaer-Larsen, J. H.; Laursen, I.; Leunbach, I.; Ehnholm, G.; Wistrand, L. G.; Petersson, J. S.; Golman, K. *Journal of Magnetic Resonance* **1998**, *133*, 1–12.

[62] Ardenkjaer-Larsen, J. H.; Macholl, S.; Johannesson, H. *Applied Magnetic Resonance* **2008**, *34*, 509–522.

[63] Hu, K.-N.; Song, C.; Yu, H.-h.; Swager, T. M.; Griffin, R. G. *J. Chem. Phys.* **2008**, *128*, 52302–17.

[64] Armstrong, B. D.; Han, S. *The Journal of Chemical Physics* **2007**, *127*, 104508.

[65] Sezer, D.; Prandolini, M. J.; Prisner, T. F. *Phys. Chem. Chem. Phys.* **2009**, *11*, 6626–6637.

[66] Denysenkov, V. P.; Prandolini, M. J.; Krahn, A.; Gafurov, M.; Endeward, B.; Prisner, T. F. *Applied Magnetic Resonance* **2008**, *34*, 289–299.

[67] Gitti, R.; Wild, C.; Tsiao, C.; Zimmer, K.; Glass, T. E.; Dorn, H. C. *Journal of the American Chemical Society* **1988**, *110*, 2294–2296.

[68] McCarney, E. R.; Han, S. *Journal of Magnetic Resonance* **2008**, *190*, 307–315.

[69] Kleschyov, A. L.; Sen, V. D.; Golubev, V. A.; Klinke, A.; Baldus, S.; Munzel, T. *Naunyn-Schmiedebergs Archives Of Pharmacology* **2008**, *377*, 277.

[70] Kleschyov, A. L.; Golubev, V. A.; Klinke, A.; Munzel, T.; Baldus, S.; Sen, V. D. *Free Radical Biology And Medicine* **2007**, *43*, S178–S178.

[71] Robinson, B. H.; Haas, D. A.; Mailer, C. *Science* **1994**, *263*, 490–493.

[72] Vasavada, K. V.; Schneider, D. J.; Freed, J. H. *J. Chem. Phys.* **1987**, *86*, 647–661.

[73] Stoll, S.; Schweiger, A. *Journal of Magnetic Resonance* **2006**, *178*, 42–55.

[74] Goldman, S. A.; Bruno, G. V.; Freed, J. H. *The Journal of Physical Chemistry* **1972**, *76*, 1858–1860.

[75] White, G.; Ottignon, L.; Georgiou, T.; Kleanthous, C.; Moore, G.; Thomson, A.; Oganesyan, V. *Journal of Magnetic Resonance* **2007**, *185*, 191–203.

[76] Hyde, J. S.; Yin, J.-J.; Subczynski, W. K.; Camenisch, T. G.; Ratke, J. J.; Froncisz, W. *The Journal of Physical Chemistry B* **2004**, *108*, 9524–9529.

[77] Froncisz, W.; Camenisch, T. G.; Ratke, J. J.; Anderson, J. R.; Subczynski, W. K.; Strangeway, R. A.; Sidabras, J. W.; Hyde, J. S. *Journal of Magnetic Resonance* **2008**, *193*, 297–304.

[78] Bennati, M.; Luchinat, C.; Parigi, G.; Turke, M.-T. *Phys. Chem. Chem. Phys.* **2010**, *12*, 5902–5910.

[79] Yamaguchi, S.; Hayashi, H.; Hamada, F.; Nakajima, A. *Polymer Bulletin* **1979**, *1*, 641–646.

[80] Dautzenberg, H. *Polyelectrolytes: Formation, Characterization and Application;* Carl Hanser Publications: Munich, 1994.

[81] Hinderberger, D.; Spiess, H.; Jeschke, G. *Applied Magnetic Resonance* **2010**, *37*, 657–683.

[82] Schild, H. G. *Progress In Polymer Science* **1992**, *17*, 163–249.

[83] Junk, M. J. N.; Jonas, U.; Hinderberger, D. *Small* **2008**, *4*, 1485–1493.

[84] Cao, Y.; Zhu, X. X.; Luo, J.; Liu, H. *Macromolecules* **2007**, *40*, 6481–6488.

[85] Engelberg, H. *Cancer* **1999**, *85*, 257–272.

[86] Mulloy, B.; Forster, M. J.; Jones, C.; Davies, D. B. *Biochem. J.* **1993**, *293*, 849–0.

[87] Atkins, E. D.; Nieduszynski, I. A. *Adv Exp Med Biol* **1975**, *52*, 19–37.

[88] Faham, S.; Hileman, R. E.; Fromm, J. R.; Linhardt, R. J.; Rees, D. C. *Science* **1996**, *271*, 1116–1120.

[89] Neville Chamberlain, L.; Edwards, I. A. S.; Stadler, H. P.; Grant Buchanan, J.; Thomas, A. *Carbohydrate Research* **1981**, *90*, 131–137.

[90] Mansson, S.; Johansson, E.; Magnusson, P.; Chai, C.-M.; Hansson, G.; Petersson, J.; Stahlberg, F.; Golman, K. *European Radiology* **2006**, *16*, 57-67.

[91] Svensson, J.; Mansson, S.; Johansson, E.; Petersson, J. S.; Olsson, L. E. *Magn. Reson. Med.* **2003**, *50*, 256–262.

[92] Johansson, E.; Olsson, L.; Mansson, S.; Petersson, J.; Golman, K.; Stahlberg, F.; Wirestam, R. *Magn. Reson. Med.* **2004**, *52*, 1043–1051.

[93] Golman, K.; Ardenkaer-Larsen, J. H.; Petersson, J. S.; Mansson, S.; Leunbach, I. *Proceedings of the National Academy of Sciences of the United States of America* **2003**, *100*, 10435–10439.

[94] Hausser, K. H. *Naturwissenschaften* **1960**, *47*, 251–251.

[95] Müller-Warmuth, W.; van Steenwinkel, R.; Yalciner, A. *Molecular Physics: An International Journal at the Interface Between Chemistry and Physics* **1971**, *21*, 449–459.

[96] Hausser, K. H.; Reinhold, F. *Zeitschrift Fur Naturforschung Part A-Astrophysik Physik Und Physikalische Chemie* **1961**, *16*, 1114–1116.

[97] Gruss, K. "Dynamische Kernspinpolarisation von 13C bei Raumtemperatur", Master's thesis, Max Planck Institute for Polymer Research, University of Mainz, 2010.

[98] Hinderberger, D.; Jeschke, G.; Spiess, H. W. *Macromolecules* **2002**, *35*, 9698–9706.

[99] Shapiro, A. B.; Baimagam, K.; Goldfeld, M. G.; Rozantse, E. G. *Zhurnal Organicheskoi Khimii* **1972**, *8*, 2263–2269.

[100] Schallenberg, E. E.; Calvin, M. *Journal Of The American Chemical Society* **1955**, *77*, 2779–2783.

[101] Maksimova, L. A.; Grigoryan, G. L.; Rozantsev, E. G. *Bulletin Of The Academy Of Sciences Of The Ussr Division Of Chemical Science* **1975**, *24*, 859–862.

[102] Fernandez, C.; Hattan, C. M.; Kerns, R. J. *Carbohydrate Research* **2006**, *341*, 1253–1265.

[103] Sen, V. D.; Golubev, V. A. *Journal Of Physical Organic Chemistry* **2009**, *22*, 138–143.

[104] Usov, A. I.; Adamyant, K. S.; Miroshni, L. I.; Shaposhn, A. A.; Kochetko, N. K. *Carbohydrate Research* **1971**, *18*, 336.

[105] Inoue, Y.; Nagasawa, K. *Carbohydrate Research* **1976**, *46*, 87–95.

[106] Huang, L. S.; Kerns, R. J. *Bioorganic & Medicinal Chemistry* **2006**, *14*, 2300–2313.

[107] Shea, K. J.; Stoddard, G. J.; Shavelle, D. M.; Wakui, F.; Choate, R. M. *Macromolecules* **1990**, *23*, 4497–4507.

[108] Toomey, R.; Freidank, D.; Ruhe, J. *Macromolecules* **2004**, *37*, 882–887.

Die VDM Verlagsservicegesellschaft sucht für wissenschaftliche Verlage abgeschlossene und herausragende

Dissertationen, Habilitationen, Diplomarbeiten, Master Theses, Magisterarbeiten usw.

für die kostenlose Publikation als Fachbuch.

Sie verfügen über eine Arbeit, die hohen inhaltlichen und formalen Ansprüchen genügt, und haben Interesse an einer honorarvergüteten Publikation?

Dann senden Sie bitte erste Informationen über sich und Ihre Arbeit per Email an *info@vdm-vsg.de*.

Sie erhalten kurzfristig unser Feedback!

VDM Verlagsservicegesellschaft mbH
Dudweiler Landstr. 99   Telefon +49 681 3720 174
D - 66123 Saarbrücken   Fax   +49 681 3720 1749
www.vdm-vsg.de

Die VDM Verlagsservicegesellschaft mbH vertritt

Printed by Books on Demand GmbH, Norderstedt / Germany